Springer Undergraduate Mathematics

Springer

London
Berlin
Heidelberg
New York
Barcelona
Budapest
Hong Kong
Milan
Paris
Santa Clara
Singapore
Tokyo

Other books in this series

Elements of Logic via Numbers and Sets
D. L. Johnson
3-540-76123-3

Basic Linear Algebra
T.S. Blyth and E.F. Robertson
3-540-76122-5

Elementary Number Theory
Gareth A. Jones and J. Mary Jones
3-540-76197-7

Introductory Mathematics: Applications and Methods
G.S. Marshall
3-540-76179-9

Vector Calculus
P.C. Matthews
3-540-76180-2

Introductory Mathematics: Algebra and Analysis
Geoff Smith
3-540-76178-0

Seán Dineen

Multivariate Calculus and Geometry

With 109 Figures

 Springer

Professor Seán Dineen
Department of Mathematics, University College Dublin,
Belfield, Dublin 4, Rupublic of Ireland

Cover illustration elements reproduced by kind permission of:
Aptech Systems, Inc., Publishers of the GAUSS Mathematical and Statistical System, 23804 S.E. Kent-Kangley Road, Maple Valley, WA 98038, USA. Tel: (206) 432 - 7855 Fax (206) 432 - 7832 email: info@aptech.com URL: www.aptech.com
American Statistical Association: Chance Vol 8 No 1, 1995 article by KS and KW Heiner 'Tree Rings of the Northern Shawangunks' page 32 fig 2
Springer-Verlag: Mathematica in Education and Research Vol 4 Issue 3 1995 article by Roman E Maeder, Beatrice Amrhein and Oliver Gloor 'Illustrated Mathematics: Visualization of Mathematical Objects' page 9 fig 11, originally published as a CD ROM 'Illustrated Mathematics' by TELOS: ISBN 0-387-14222-3, german edition by Birkhauser: ISBN 3-7643-5100-4.
Mathematica in Education and Research Vol 4 Issue 3 1995 article by Richard J Gaylord and Kazume Nishidate 'Traffic Engineering with Cellular Automata' page 35 fig 2. Mathematica in Education and Research Vol 5 Issue 2 1996 article by Michael Trott 'The Implicitization of a Trefoil Knot' page 14.
Mathematica in Education and Research Vol 5 Issue 2 1996 article by Lee de Cola 'Coins, Trees, Bars and Bells: Simulation of the Binomial Process page 19 fig 3. Mathematica in Education and Research Vol 5 Issue 2 1996 article by Richard Gaylord and Kazume Nishidate 'Contagious Spreading' page 33 fig 1. Mathematica in Education and Research Vol 5 Issue 2 1996 article by Joe Buhler and Stan Wagon 'Secrets of the Madelung Constant' page 50 fig 1.

ISBN 3-540-76176-4 Springer-Verlag Berlin Heidelberg New York

British Library Cataloguing in Publication Data
Dineen, Seán, 1944 -
    Multivariate calculus and geometry. - (Springer undergraduate mathematics series)
    1. Calculus   2. Multivariate analysis   3. Geometry
    I. Title
    515
ISBN 3540761764

Library of Congress Cataloging-in-Publication Data
Dineen, Seán 1944 -
    Multivariate calculus and geometry / Seán Dineen.
        p.   cm. -- (Springer undergraduate mathematics series)
    Includes index.
    ISBN 3-540-76176-4 (alk. paper)
    1. Calculus I. Title II. Series
QA303.D63 1998   97-31752
515--dc21   CIP

Typeset by Focal Image, London
Printed and bound at the Athenæum Press Ltd., Gateshead, Tyne & Wear
12/3830-54321 Printed on acid-free paper

To my four godchildren

Anne-Marie Dineen, Donal Coffey, Kevin Timoney, Eoghan Wallace.

Contents

Preface

The importance assigned to accuracy in basic mathematics courses has, initially, a useful disciplinary purpose but can, eventually, hinder progress if it fosters the belief that exactness is all that makes mathematics what it is. *Multivariate calculus* occupies a pivotal position in undergraduate mathematics programmes in providing students with the opportunity to outgrow this narrow viewpoint and to develop a flexible, intuitive and independent vision of mathematics. This possibility arises from the extensive nature of the subject.

Multivariate calculus links together in a non-trivial way, perhaps for the first time in a student's experience, four important subject areas: *analysis*, *linear algebra*, *geometry* and *differential calculus*. Important features of the subject are reflected in the variety of alternative titles we could have chosen, e.g. "Advanced Calculus", "Vector Calculus", "Multivariate Calculus", "Vector Geometry", "Curves and Surfaces" and "Introduction to Differential Geometry". Each of these titles partially reflects our interest but it is more illuminating to say that here we study differentiable functions, i.e.

functions which enjoy a good local approximation by linear functions.

The main emphasis of our presentation is on understanding the underlying fundamental principles. These are discussed at length, carefully examined in simple familiar situations and tested in technically demanding examples. This leads to a structured and systematic approach of manageable proportions which gives shape and coherence to the subject and results in a comprehensive and unified exposition.

We now discuss the four underlying topics and the background we expect – bearing in mind that the subject can be approached with different levels of mathematical maturity. Results from *analysis* are required to justify much of this book yet many students have little or no background in analysis when they approach multivariate calculus. This is not surprising as differential calculus preceded and indeed motivated the development of analysis. We do not list analysis as a prerequisite but hope that our presentation shows its importance and motivates the reader to study it further.

Since linear approximations appear in the definition of differentiable functions it is not surprising that *linear algebra* plays a part in this book. Several-variable calculus and linear algebra developed, to a certain extent, side by side to their mutual benefit. The primary role of linear algebra, in our study, is to provide a suitable notation and framework in which we can clearly and compactly introduce concepts and present and prove results. This is more important than it appears since to quote T.C. Chaundy "notation biases analysis as language biases thought". An elementary knowledge of matrices and determinants is assumed and particular results from linear algebra are introduced as required.

We discuss the role of *geometry* in multivariate calculus throughout the text and confine ourselves here to a brief comment. The natural setting for *functions* which enjoy a good local approximation by *linear functions* are *sets* which enjoy a good local approximation to *linear spaces*. In one and two dimensions this leads to *curves* and *surfaces*, respectively, and in higher dimensions to *differentiable manifolds*.

We assume the reader has acquired a reasonable knowledge of *one-variable differential and integral calculus* before approaching this book. Although not assumed, some experience with partial derivatives allows the reader to proceed rapidly through routine calculations and to concentrate on important concepts. A reader with no such experience should definitely read Chapter 1 a few times before proceeding and may even wish to consult the author's *Functions of Two Variables* (Chapman and Hall, 1995).

We now turn to the contents of this book. Our general approach is holistic and we hope that the reader will be equally interested in all parts of this book. Nevertheless, it is possible to group certain chapters thematically.

Differential Calculus on Open Sets and Surfaces (Chapters 1–4).
We discuss extremal values of real-valued functions on surfaces and open sets. The important principle here is the *Implicit Function Theorem* which links linear approximations with systems of linear equations and sets up a relationship between graphs and surfaces.

Integration Theory (Chapters 6, 9, 11–15).
The key concepts are parametrizations (Chapters 5, 10 and 14) and oriented

surfaces (Chapter 12). We build up our understanding and technical skill step by step, by discussing in turn line integrals (Chapter 6), integration over open subsets of $\mathbb{R}^2$ (Chapter 9), integration over simple surfaces without orientation (Chapter 11), integration over simple oriented surfaces (Chapter 12) and triple integrals over open subsets of $\mathbb{R}^3$ (Chapter 14). At the appropriate times we discuss generalizations of the fundamental theorem of calculus, i.e. *Green's Theorem* (Chapter 9), *Stokes' Theorem* (Chapter 13) and *the Divergence Theorem* (Chapter 15). Special attention is given to the parametrization of classical surfaces, the evaluation of surface integrals using projections, the change of variables formula and to the detailed examination of involved geometric examples.

Geometry of Curves and Surfaces (Chapters 5, 7–8, 10, 16–18).
We discuss signed curvature in $\mathbb{R}^2$ and use vector-valued differentiation to obtain the Frenet–Serret equations for curves in $\mathbb{R}^3$. The abstract geometric study of surfaces using Gaussian curvature is, regrettably, usually not covered in multivariate calculus courses. The fundamental concepts, parametrizations and plane curvature, are already in place (Chapters 5, 7 and 10) and examples from integration theory (Chapters 11–15) provide a concrete background and the required geometric insight. Using only curves in $\mathbb{R}^2$ and critical points of functions of two variables we develop the concept of Gaussian curvature. In addition, we discuss normal, geodesic and intrinsic curvature and establish a relationship between all three. In the final chapter we survey informally a number of interesting results from differential geometry.

This text is based on a course given by the author at University College Dublin. The additions that emerged in providing details and arranging self-sufficiency suggest that it is suitable for a course of 30 lectures. Although the different topics come together to form a unified subject, with different chapters mutually supporting one another, we have structured this book so that each chapter is self-contained and devoted to a single theme.

This book can be used as a main text, as a supplementary text or for self-study. The groupings summarised above allow a selection of short courses at a slower pace. The exercises are extremely important as it is through them that a student can assess progress and understanding.

Our aim was to write a short book focusing on basic principles while acquiring technical skills. This precluded comments on the important applications of multivariate calculus which arise in physics, statistics, engineering, economics and, indeed, in most subjects with scientific aspirations.

It is a pleasure to acknowledge the help I received in bringing this project to fruition. Dana Nicolau displayed skill in preparing the text and great patience in accepting with a cheerful "OK, OK," the continuous stream of revisions, corrections and changes that flowed her way. Michael Mackey's diagrams speak for themselves. Brendan Quigley's geometric insight and Pauline Mellon's suggestions helped shape our outlook and the text. I would like especially to thank the Third Arts students at University College Dublin, and especially Tim Cronin and Martin Brundin for their comments, reactions and corrections. Susan Hezlet of Springer provided instantaneous support, ample encouragement and helpful suggestions at all times. To all these and the community of mathematicians whose results and writings have influenced me I say – thank you!

Department of Mathematics,
University College Dublin,
Belfield, Dublin 4,
Ireland.

1

Introduction to Differentiable Functions

Summary. *We introduce differentiable functions, directional and partial derivatives, graphs and level sets of functions of several variables.*

If, with little or no difficulty, you can integrate xe^x over $[0, 1]$, invert a 3×3 matrix and calculate the partial derivatives of $\sin(xy + y^2)$ then you probably have the prerequisites needed to read this book. A formal summary of the background required is more complicated since the underlying topics have to be treated separately.

The logical foundation for a number of results is Theorem 1.1 and this comes to us from analysis. Indeed, analysis is the shadow that precedes everything in this book. We do not assume familiarity with this subject but hope that our approach will motivate the reader to undertake a serious study of analysis after reading this book. Differentiable functions are precisely those functions which enjoy a good linear approximation and hence it is no surprise that linear algebra plays a role in our study. This topic, which we develop as we proceed, supplies us with a number of useful concepts and results as well as a notation and language in which we can efficiently frame results and explain proofs.

It goes without saying that we assume a knowledge of one-variable differential and integral calculus. The main topic in this chapter is the introduction, in a rather concise fashion, of differentiable functions and associated concepts, e.g. partial derivatives, directional derivatives, higher derivatives, and their presentation in the format that we require. Although not essential, some familiarity with partial derivatives would help focus the reader's attention on the more essential aspects of the topics discussed. The reader, without such a background, is advised to read this rather condensed chapter a few times before proceeding and strongly advised to write out in full detail solutions to Exercises 1.2–1.5.

We begin our study by introducing some notation. Appropriate notation is often the difference between simple and complicated presentations of several-variable calculus. Usually, but not always, we use capital letters or boldface to denote vectors and vector-valued functions and lower case letters to denote real numbers and real-valued functions.

Let $X = (x_1, \ldots, x_n)$ denote a typical point in $\mathbb{R}^n$. The *length* (or *norm*) of X, $\|X\|$, is defined as $(x_1^2 + \cdots + x_n^2)^{1/2}$. If X and $Y = (y_1, \ldots, y_n)$ are vectors in $\mathbb{R}^n$ then the *inner product* (or *dot product* or *scalar product*) of X and Y, $X \cdot Y$ or $\langle X, Y \rangle$, is defined as

$$\langle X, Y \rangle = X \cdot Y = \sum_{i=1}^{n} x_i y_i.$$

We have $\|X\|^2 = \langle X, X \rangle$ and two vectors X and Y are *perpendicular* if and only if their inner product is zero.

A subset U of $\mathbb{R}^n$ is *open* if for each $X_0 \in U$ there exists $\varepsilon > 0$ such that

$$\{X \in \mathbb{R}^n : \|X - X_0\| < \varepsilon\} \subset U.$$

A subset A of $\mathbb{R}^n$ is *closed* if its complement is open and a set B is *bounded* if there exists $M \in \mathbb{R}$ such that $\|x\| \leq M$ for all $x \in B$. A crucial role in all aspects of calculus, analysis and geometry is played by sets which are both closed and bounded; such sets are said to be *compact*.

Example 1.1

The closed solid sphere $\{(x, y, z) : x^2 + y^2 + z^2 \leq 1\}$ and its boundary $\{(x, y, z) : x^2 + y^2 + z^2 = 1\}$ are both compact subsets of $\mathbb{R}^3$ while the solid sphere without boundary $\{(x, y, z) : x^2 + y^2 + z^2 < 1\}$ is an open subset of $\mathbb{R}^3$. Any open subset of $\mathbb{R}^3$ is a union of open spheres.

A function $F: U \subset \mathbb{R}^n \to \mathbb{R}^m$ is *continuous* if for each $X_0 \in U$ and each $\varepsilon > 0$ there exists $\delta > 0$ such that

$$\|F(X) - F(X_0)\| \leq \varepsilon$$

whenever $X \in U$ and $\|X - X_0\| \leq \delta$. Continuous functions may also be defined by using convergent sequences (see Exercises 1.16 and 1.17).

A function $f: A \subset \mathbb{R}^n \to \mathbb{R}$ has a *maximum* on A if there exists $X_1 \in A$ such that $f(X) \leq f(X_1)$ for all $X \in A$. We call $f(X_1)$ the maximum of f on A and say that f achieves its maximum on A at X_1. The maximum, if it exists, is unique but may be achieved at more than one point. A point X_1 in A is called a *local maximum* of f on A if there exists $\delta > 0$ such that f achieves its maximum on $A \cap \{X : \|X - X_1\| < \delta\}$ at X_1. If, in addition, $f(X) < f(X_1)$ whenever $X \neq X_1$ we call X_1 a *strict local maximum*. Isolated local maxima are strict, i.e. if for some $\delta > 0$, X_1 is the only local maximum of f in $A \cap \{X : \|X - X_1\| < \delta\}$ then X_1 is a strict local maximum. In particular, if the set of local maxima of f is *finite* then all local maxima are strict local maxima. The analogous definitions of minimum, local minimum and strict local minimum are obtained by reversing the above inequalities.

Compact sets and continuity feature in the following *fundamental existence theorem for maxima and minima.*

Theorem 1.1

A continuous real-valued function on a compact subset of $\mathbb{R}^n$ has a maximum and a minimum.

The practical problem of finding maxima and minima often requires a degree of smoothness finer than continuity called differentiability.

Definition 1.1

Let $F : U \subset \mathbb{R}^n \to \mathbb{R}^m$, U open, and let $X \in U$. We say that F is differentiable at X if there exists an $m \times n$ matrix A, $\varepsilon > 0$ and a function

$$G : \{h \in \mathbb{R}^n : \|h\| < \varepsilon\} \longrightarrow \mathbb{R}^m$$

such that

$$F(X + h) = F(X) + A \circ h + G(h)\|h\|$$

for $\|h\| < \varepsilon$ where $G(h) \to 0$ as $h \to 0$.

The matrix A is necessarily unique and written $DF(X)$ or $F'(X)$ and called the *derivative* of F at X. Note that if $F : \mathbb{R}^n \to \mathbb{R}^m$ then $F'(X)$ is an $m \times n$ matrix (i.e. the order is reversed). The term $F(X) + F'(X) \circ h$ is our *linear approximation* to $F(X + h)$ and $\|G(h)h\|$ is the *error* in our approximation. Note that $h = (h_1, \ldots, h_n)$ and $F'(X) \circ h$ is matrix multiplication of the $m \times n$ matrix $F'(X)$ and the $n \times 1$ matrix h.

If F is differentiable then F is continuous. The composition of differentiable functions is again differentiable and the *chain rule*, which gives the derivative of the composition, is most elegantly phrased using matrix notation. Let $F : U$ (open) $\subset \mathbb{R}^n \to \mathbb{R}^m$ and $G : V$ (open) $\subset \mathbb{R}^m \to \mathbb{R}^p$ denote differentiable functions. If $P \in U$ and $F(P) \in V$ then $(G \circ F)'(P)$ exists and the *chain rule* tells us that

$$(G \circ F)'(P) = G'(F(P)) \circ F'(P).$$

Continuity and differentiability of most functions we encounter can be verified by using functions from $\mathbb{R}$ into $\mathbb{R}$, addition, multiplication and composition of functions and the *coordinate functions* that we now describe. For $1 \leq i \leq n$, let $\mathbf{e}_i = (0, \ldots, 1, 0, \ldots)$, where 1 lies in the ith position. The set $(\mathbf{e}_i)_{i=1}^n$ is a *basis* for the *vector space* $\mathbb{R}^n$ (and is often called the *standard unit vector basis* for $\mathbb{R}^n$). If $X = (x_1, \ldots, x_n) \in \mathbb{R}^n$ then

$$X = x_1\mathbf{e}_1 + x_2\mathbf{e}_2 + \cdots + x_n\mathbf{e}_n = \sum_{i=1}^n x_i\mathbf{e}_i.$$

We define $\mathbf{e}_i^*$, $1 \leq i \leq n$, the *coordinate functions* from $\mathbb{R}^n$ into $\mathbb{R}$, by

$$\mathbf{e}_i^*(X) = \mathbf{e}_i^*(x_1, \ldots, x_n) = \langle \mathbf{e}_i, X \rangle = x_i.$$

The set $(\mathbf{e}_i^*)_{i=1}^n$ is a basis for $(\mathbb{R}^n)^*$ – the vector space of all real-valued linear functions on $\mathbb{R}^n$. It is called the *dual basis* to $(\mathbf{e}_i)_{i=1}^n$ since

$$\mathbf{e}_i^*(\mathbf{e}_j) = \begin{cases} 1 & \text{if } i = j \\ 0 & \text{if } i \neq j. \end{cases}$$

When $n = 3$ we also write (x, y, z) to denote a typical point in $\mathbb{R}^3$ and sometimes use $\mathbf{i}$, $\mathbf{j}$ and $\mathbf{k}$ in place of $\mathbf{e}_1$, $\mathbf{e}_2$ and $\mathbf{e}_3$ respectively.

Let U denote an open subset of $\mathbb{R}^n$ and let $F: U \to \mathbb{R}^m$. If $X \in U$ then $F(X)$ has m coordinates which we denote by $f_1(X), \ldots, f_m(X)$. Thus we often write $F = (f_1, \ldots, f_m)$ where each f_i is a real-valued function of n variables. If $\mathbf{v} \in \mathbb{R}^n$ we let

$$D_{\mathbf{v}}F(X) = \lim_{\substack{t \to 0 \\ t \in \mathbb{R}}} \frac{F(X + t\mathbf{v}) - F(X)}{t}$$

and call $D_{\mathbf{v}}F(X)$ the *directional derivative* of F in the direction $\mathbf{v}$. This is the derivative of the mapping $t \in \mathbb{R} \to F(X + t\mathbf{v}) \in \mathbb{R}^m$ and is an $m \times 1$ matrix (although it is often written incorrectly, for typographical reasons, as a $1 \times m$ row matrix).

If $\mathbf{v} = \mathbf{e}_i$ we write $\dfrac{\partial F}{\partial x_i}(X)$ or $F_{x_i}(X)$ in place of $D_{\mathbf{e}_i}F(X)$ and call these the (first-order) *partial derivatives* of F. We let $\nabla f_i(X) = \left(\dfrac{\partial f_i}{\partial x_1}(X), \ldots, \dfrac{\partial f_i}{\partial x_n}(X) \right)$ for all i and call $\nabla f_i(X)$ the *gradient* of f_i at X.

If F is differentiable at X then all (first-order) directional and partial derivatives of F exist and

$$F'(X) = \begin{pmatrix} \nabla f_1(X) \\ \vdots \\ \nabla f_m(X) \end{pmatrix} = \begin{pmatrix} \dfrac{\partial f_1}{\partial x_1}(X) & \cdots & \dfrac{\partial f_1}{\partial x_n}(X) \\ \vdots & & \vdots \\ \dfrac{\partial f_m}{\partial x_1}(X) & \cdots & \dfrac{\partial f_m}{\partial x_n}(X) \end{pmatrix} = \left(F_{x_1}(X), \ldots, F_{x_n}(X) \right).$$

Moreover, if $\mathbf{v} = (v_1, \ldots, v_n) \in \mathbb{R}^n$, then

$$D_{\mathbf{v}} F(X) = F'(X) \circ \mathbf{v} = \sum_{i=1}^{n} v_i \frac{\partial F}{\partial x_i}(X).$$

Higher-order directional and partial derivatives are defined in the usual way, i.e. by repeated differentiation.

Example 1.2

Let $F: \mathbb{R}^4 \to \mathbb{R}^3$ be defined by

$$F(x, y, z, w) = (x^2 y, xyz, x^2 + y^2 + zw^2).$$

Then $F = (f_1, f_2, f_3)$ where $f_1(x, y, z, w) = x^2 y$, $f_2(x, y, z, w) = xyz$ and $f_3(x, y, z, w) = x^2 + y^2 + zw^2$. Moreover, $\nabla f_1(x, y, z, w) = (2xy, x^2, 0, 0)$, $\nabla f_2(x, y, z, w) = (yz, xz, xy, 0)$ and $\nabla f_3(x, y, z, w) = (2x, 2y, w^2, 2zw)$. Hence

$$F'(x, y, z, w) = \begin{pmatrix} 2xy & x^2 & 0 & 0 \\ yz & xz & xy & 0 \\ 2x & 2y & w^2 & 2zw \end{pmatrix}.$$

If $X = (1, 2, -1, -2)$ and $\mathbf{v} = (0, 1, 2, -2)$ then

$$D_{\mathbf{v}} F(X) = F'(X) \circ \mathbf{v} = \begin{pmatrix} 4 & 1 & 0 & 0 \\ -2 & -1 & 2 & 0 \\ 2 & 4 & 4 & 4 \end{pmatrix} \begin{pmatrix} 0 \\ 1 \\ 2 \\ -2 \end{pmatrix} = \begin{pmatrix} 1 \\ 3 \\ 4 \end{pmatrix}.$$

If all first-order partial derivatives of F exist and are continuous then F is differentiable. This criterion enables us to see, literally at a glance, the sets on which most functions have derivatives, partial derivatives and directional derivatives. Consider, for instance, the function

$$F(x, y, z) = (\sin(xyz), x^2 - y^2, \exp(xy)).$$

We have

$$F(x, y, z) = \sin(xyz)\mathbf{e}_1 + (x^2 - y^2)\mathbf{e}_2 + \exp(xy)\mathbf{e}_3$$

and it suffices to consider separately the three $\mathbb{R}$-valued functions $\sin(xyz)$, $x^2 - y^2$ and $\exp(xy)$. Since $\mathbf{e}_1^*(x, y, z) = x$, $\mathbf{e}_2^*(x, y, z) = y$ and $\mathbf{e}_3^*(x, y, z) = z$ we have

$$\left(\sin(\mathbf{e}_1^* \mathbf{e}_2^* \mathbf{e}_3^*)\right)(x, y, z) = \sin\left(\mathbf{e}_1^*(x, y, z)\mathbf{e}_2^*(x, y, z)\mathbf{e}_3^*(x, y, z)\right) = \sin(xyz)$$

and, after similar calculation for $x^2 - y^2$ and $\exp(xy)$, we have

$$F = \big(\sin(\mathbf{e}_1^* \mathbf{e}_2^* \mathbf{e}_3^*)\big)\mathbf{e}_1 + \big((\mathbf{e}_1^*)^2 - (\mathbf{e}_2^*)^2\big)\mathbf{e}_2 + \big(\exp(\mathbf{e}_1^* \mathbf{e}_2^*)\big)\mathbf{e}_3 \,.$$

We are reduced to considering properties of the functions $\mathbf{e}_1^*$, $\mathbf{e}_2^*$, $\mathbf{e}_3^*$ and functions from $\mathbb{R}$ into $\mathbb{R}$ such as $\sin$ and $\exp$. Since

$$\frac{\partial \mathbf{e}_i^*}{\partial x_j}(X) = \begin{cases} 1 & \text{if } i = j \\ 0 & \text{if } i \neq j \end{cases}$$

for all X we see that $\mathbf{e}_i^*$ has continuous partial derivatives of *all orders*. Hence F is differentiable and has continuous partial and directional derivatives of all orders on $\mathbb{R}^3$.

Associated with any function $F \colon U \subset \mathbb{R}^n \to \mathbb{R}^m$ we have two types of sets which play a special role in the development of the subject – graphs and level sets. The *graph* of F is the subset of $\mathbb{R}^{n+m}$ defined as follows

$$\begin{aligned} \text{graph}\,(F) \;&= \big\{(X, Y) : X \in U \text{ and } Y = F(X)\big\} \\ &= \big\{(X, F(X)) : X \in U\big\}. \end{aligned}$$

If $C = (c_1, \ldots, c_m)$ is a point in $\mathbb{R}^m$ we define the *level set* of F, $F^{-1}(C)$, by the formula

$$F^{-1}(C) = \big\{X \in U : F(X) = C\big\}.$$

In terms of the coordinate functions $f_1, \ldots, f_m$ of F we have

$$F(X) = C \iff f_i(X) = c_i \text{ for } i = 1, \ldots, m$$

and hence

$$F^{-1}(C) = \bigcap_{i=1}^{m} \big\{X \in U : f_i(X) = c_i\big\} = \bigcap_{i=1}^{m} f_i^{-1}(c_i).$$

Thus a level set of a vector-valued function is the finite intersection of level sets of real-valued functions. This is frequently useful in arriving at a geometrical interpretation of level sets as the following example shows.

Example 1.3

Let $F(x, y, z) = (x^2 + y^2 + z^2 - 1, 2x^2 + 2y^2 - 1)$. We have $F = (f_1, f_2)$ where $f_1(x, y, z) = x^2 + y^2 + z^2 - 1$ and $f_2(x, y, z) = 2x^2 + 2y^2 - 1$. The set $f_1^{-1}(0)$ is a *sphere* of radius 1 while $f_2^{-1}(0)$ is a *cylinder* parallel to the z–axis built on a circle with centre the origin and radius $1/\sqrt{2}$. If $\mathbf{0} = (0, 0)$ is the origin in $\mathbb{R}^2$ then

$$F^{-1}(\mathbf{0}) = f_1^{-1}(0) \cap f_2^{-1}(0)$$

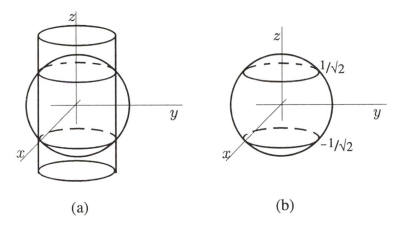

(a)   (b)

Figure 1.1.

is the intersection of a sphere and a cylinder in $\mathbb{R}^3$ (Figure 1.1(a)).

For this particular example we obtain more information by solving the equations $f_1(x, y, z) = f_2(x, y, z) = 0$. We have $x^2 + y^2 = 1 - z^2 = 1/2$. Hence $z^2 = 1/2$, $z = \pm 1/\sqrt{2}$ and the level set consists of two circles on the unit sphere (Figure 1.1(b)).

The relationship between graphs and level sets plays an important role in our study. The easy part of this relationship – every graph is a level set – is given in the next example while the difficult part – every (regular) level set is locally a graph – is the *implicit function theorem* (Chapter 2).

Example 1.4

Let $F: U \subset \mathbb{R}^n \to \mathbb{R}^m$. We define $G: U \times \mathbb{R}^m \to \mathbb{R}^m$ by $G(X, Y) = F(X) - Y$. If $\mathbf{0}$ is the origin in $\mathbb{R}^m$ then

$$
\begin{aligned}
(X, Y) \in G^{-1}(\mathbf{0}) \quad &\Longleftrightarrow \quad G(X, Y) = \mathbf{0} \\
&\Longleftrightarrow \quad F(X) - Y = \mathbf{0} \\
&\Longleftrightarrow \quad (X, Y) \in \text{graph}\,(F).
\end{aligned}
$$

Hence $G^{-1}(\mathbf{0}) = \text{graph}\,(F)$ and every graph is a level set.

EXERCISES

1.1 Sketch the following sets

(a) $\left\{(x, y, z) : \dfrac{x^2}{a^2} + \dfrac{y^2}{b^2} + \dfrac{z^2}{c^2} < 1\right\}$

(b) $\left\{(x, y, z) : x^2 + y^2 = z^2\right\}$

(c) $\left\{(x, y, z) : x \geq 0,\ y \geq 0,\ z \geq 0\right\}$

(d) $\left\{(x, y, z) : x^2 + y^2 + z^2 - 2z = 0\right\}$

(e) $\left\{(x, y, z) : x^2 + y^2 + z^2 - 4z = 0,\ x^2 + y^2 = 4\right\}$.

From your sketches determine which of the sets are: open, closed, bounded, compact.

1.2 Find all first-order partial derivatives of

(a) $f(x, y, z) = (z^2 + x^2) \log(1 + x^2 y^2)$

(b) $g(x, y, z) = xy \tan^{-1}(xz)$

(c) $h(x, y, z) = \dfrac{\sin(x^2 + y^2 + z^2 + w^2)}{1 + (x - y)^2}$

1.3 If $F(x, y, z, w) = (x^2 - y^2, 2xy, zx, z^2 w^2 x^2)$ and $\mathbf{v} = (2, 1, -2, -1)$ find $F'(1, 2, -1, -2)$ and $D_{\mathbf{v}} F(1, 2, -1, -2)$.

1.4 Let $f(x, y, z) = x^2 - xy + yz^3 - 6z$. Find all points (x, y, z) such that $\nabla f(x, y, z) = (0, 0, 0)$.

1.5 If $f(x, y, z) = x^2 e^y$ and $g(x, y, z) = y^2 e^{xz}$ find ∇f, ∇g and $\nabla(fg)$. Verify that $\nabla(fg) = f \nabla g + g \nabla f$.

1.6 Let $P\colon (a, b) \subset \mathbb{R} \to \mathbb{R}^n$. Show that

$$\frac{d}{dt}\left(\|P(t)\|^2\right) = 2P'(t) \cdot P(t)$$

and deduce that if $\|P(t)\|$ does not depend on t then $P'(t) \perp P(t)$.

1.7 If $F(x, y, z) = (x^2, y^2 + z^2, xyz)$ and $G(x, y, z) = (e^x, y^2 - z^2, xyz)$ find $F'(x, y, z)$ and $G'(x, y, z)$. Let $H(x, y, z) = \langle F(x, y, z), G(x, y, z) \rangle$ where $\langle\ ,\ \rangle$ is the inner product in $\mathbb{R}^3$. Find $\nabla H(x, y, z)$ and verify that

$$\nabla H(x, y, z) = G(x, y, z) \circ F'(x, y, z) + F(x, y, z) \circ G'(x, y, z).$$

1.8 If $f(x, y, z) = (x^2 + y^2 + z^2)^{-1/2}$ show that

$$\nabla f(x, y, z) = -\frac{(x\mathbf{i} + y\mathbf{j} + z\mathbf{k})}{(x^2 + y^2 + z^2)^{3/2}}.$$

1.9 If $F: U \subset \mathbb{R}^n \to \mathbb{R}^m$ is differentiable and $F(P) \neq 0$ show that $\|F\|$ is differentiable at the point P. If $\mathbf{v} \in \mathbb{R}^n$ show that

$$\nabla_{\mathbf{v}}(\|F\|)(P) = \frac{\langle D_{\mathbf{v}}F, F \rangle}{\|F\|}(P).$$

1.10 Use Exercise 1.9 to give another proof of Exercise 1.8, i.e. first find $D_{\mathbf{e_i}}\left(\frac{1}{\|x\|}\right)$ for $i = 1, 2, 3$.

1.11 If

$$(x, y, z) \xrightarrow{F} (xyz, x^2 + y^2, x^2 - y^2, z^2)$$

$$\|$$

$$(u, v, w, t) \xrightarrow{G} (u^2 + v^2, u^2 - v^2, w^2 - t^2, w^2 + t^2)$$

$G = (G_1, G_2, G_3, G_4)$ and $H = (H_1, H_2, H_3, H_4) = G \circ F$ verify that

$$\frac{\partial H_2}{\partial x} = \frac{\partial G_2}{\partial u} \cdot \frac{\partial u}{\partial x} + \frac{\partial G_2}{\partial v} \cdot \frac{\partial v}{\partial x} + \frac{\partial G_2}{\partial w} \cdot \frac{\partial w}{\partial x} + \frac{\partial G_2}{\partial t} \cdot \frac{\partial t}{\partial x}$$

directly and also by using $H' = G' \circ F'$.

1.12 If the function $f: \mathbb{R}^n \to \mathbb{R}$ satisfies

$$\sum_{i=1}^{n} x_i^2 \frac{\partial^2 f}{\partial x_i^2} + \sum_{i=1}^{n} x_i \frac{\partial f}{\partial x_i} = 0$$

show that $h(x_1, \ldots, x_n) = f(e^{x_1}, \ldots, e^{x_n})$ satisfies

$$\sum_{i=1}^{n} \frac{\partial^2 h}{\partial x_i^2} = 0.$$

1.13 Let $f: \mathbb{R}^n \to \mathbb{R}$ be differentiable, let $P \in \mathbb{R}^n$ and $\Delta X = (\Delta x_1, \ldots, \Delta x_n) \in \mathbb{R}^n$. Show that there exists $g: \mathbb{R}^{2n} \to \mathbb{R}$ such that

$$f(P + \Delta X) = f(P) + \sum_{i=1}^{n} \frac{\partial f}{\partial x_i}(P)\Delta x_i + g(P, \Delta X)\|\Delta X\|$$

and $g(P, \Delta X) \to 0$ as $\Delta X \to 0$.

1.14 Let $f(x,y,z) = x^2y^2 + y^2z^2 + xyz$. By using the previous exercise and the values of f and its first-order derivatives at $(1,1,1)$ estimate $f(1.1, 1.05, 0.95)$. Find the error in your approximation and the error as a percentage of $f(1,1,1)$.

1.15 Identify geometrically and sketch the level set $F^{-1}(C)$ where $F(x,y,z) = (z^2 - x^2 - y^2, 2x - y)$ and $C = (1,2)$.

1.16 Let $X_j = (x_1^j, \ldots, x_n^j) \in \mathbb{R}^n$. Show that $X_j \to X$ as $j \to \infty$ where $X = (x_1, \ldots, x_n)$ if and only if $x_i^j \to x_i$ as $j \to \infty$ for all i.

1.17 Let $F = (f_1, \ldots, f_m) \in \mathbb{R}^n \to \mathbb{R}^m$. Show that the following conditions are equivalent

(a) F is continuous

(b) each f_i is continuous

(c) if $X_j \in \mathbb{R}^n \to X \in \mathbb{R}^n$ as $j \to \infty$ then $f_i(X_j) \to f_i(X)$ as $j \to \infty$ for each i.

1.18 Show that the sum, product and quotient of continuous (respectively differentiable) functions are continuous (respectively differentiable) and that differentiable functions are continuous.

1.19 If $F: U \subset \mathbb{R}^n \to \mathbb{R}^m$, $P \in U$ and L is a linear function from $\mathbb{R}^n$ to $\mathbb{R}^m$ satisfying

$$F(P + X) = F(P) + L(X) + G(X) \cdot \|X\|$$

for X close to zero and $G(X) \to 0$ as $X \to 0$ show that F is differentiable at P and $F'(X) = L$.

1.20 Let $F(x_1, x_2, x_3, x_4) = \left(\sin(x_1x_2x_3), e^{-x_1^2 + x_4^2}\right)$ for $(x_1, x_2, x_3, x_4) \in \mathbb{R}^4$. Show that

$$F = \left(\sin(e_1^*e_2^*e_3^*)\right)e_1 + \left(\exp(-(e_1^*)^2 + (e_4^*)^2)\right)e_2$$

and hence deduce that F is continuous and differentiable.

2

Level Sets and Tangent Spaces

Summary. *Using systems of linear equations as a guide we discuss the significance of the implicit function theorem for level sets. We define the tangent space and the normal space at a point on a level set.*

We shall be concerned with many different aspects of surfaces, level sets and graphs in this book. In this chapter we consider the role of differentiability in the local structure of level sets. By considering the linear approximation of differentiable functions and standard results on solving systems of linear equations we begin to appreciate and accept that level sets are locally graphs.

Let $F: U \subset \mathbb{R}^n \to \mathbb{R}^m$, U open, $F = (f_1, \ldots, f_m)$, $C = (c_1, \ldots, c_m) \in \mathbb{R}^m$. We suppose that F is differentiable and consider the level set $F^{-1}(C) = \bigcap_{i=1}^m f_i^{-1}(c_i)$, i.e. the points $(x_1, \ldots, x_n) \in U$ which satisfy the equations

$$
\begin{aligned}
f_1(x_1, \ldots, x_n) &= c_1 \\
&\;\;\vdots \\
f_m(x_1, \ldots, x_n) &= c_m.
\end{aligned}
\tag{2.1}
$$

We have n unknowns, $x_1, \ldots, x_n$ and m equations. If each f_i is linear we have a *system of linear equations* and the rank of the matrix of coefficients gives information on the number of linearly independent solutions and procedures on how to identify a complete set of independent variables. The *implicit function theorem* says that this process is also valid *locally* for differentiable functions. The key, of course, is the fact that differentiable functions, by definition, enjoy a good local *linear* approximation.

11

If $P \in F^{-1}(C)$ then $F(P) = C$. If $X \in \mathbb{R}^n$ is close to zero then, since F is differentiable,
$$F(P + X) = F(P) + F'(P) \circ X + G(X).$$
If we wish to find X close to 0 such that $F(P+X) = C$ then we are considering points such that
$$F'(P) \circ X + G(X) = 0$$
and hence $F'(P) \circ X \approx 0$ (where $\approx$ denotes approximately equal).

We assume from now on that $n \geq m$. We thus have something very close to the following system of *linear* equations

$$\frac{\partial f_1}{\partial x_1}(P)x_1 + \quad \cdots + \frac{\partial f_1}{\partial x_n}(P)x_n = 0$$
$$\vdots \qquad\qquad\qquad\qquad (2.2)$$
$$\frac{\partial f_m}{\partial x_1}(P)x_1 + \quad \cdots + \frac{\partial f_m}{\partial x_n}(P)x_n = 0.$$

The matrix of coefficients of this system of linear equations is

$$\left(\frac{\partial f_i}{\partial x_j}(P) \right)_{\substack{1 \leq i \leq m \\ 1 \leq j \leq n}}$$

i.e. $F'(P)$. From linear algebra we have

$$\text{Rank}\left(F'(P)\right) = m \iff \text{the } m \text{ rows of } F'(P) \text{ are linearly independent}$$
$$\iff \text{there exist } m \text{ linearly independent columns}$$
$$\text{in } F'(P)$$
$$\iff F'(P) \text{ contains } m \text{ columns such that the}$$
$$\text{associated } m \times m \text{ matrix has non-zero}$$
$$\text{determinant}$$
$$\iff \text{the space of solutions for the system (2.2)}$$
$$\text{is } n - m \text{ dimensional.}$$

Moreover, if any of these conditions are satisfied and we choose m columns which are linearly independent then the variables corresponding to the *remaining columns* can be taken as a *complete set of independent variables*. If the above conditions are satisfied we say that F has *full* or *maximum rank* at P.

As a simple example consider

$$2x - y + z \qquad = 0$$
$$y \qquad - w = 0$$

with matrix of coefficients

$$A = \begin{pmatrix} 2 & -1 & 1 & 0 \\ 0 & 1 & 0 & -1 \end{pmatrix}.$$

The 2×2 matrix obtained by using the first two columns is $\begin{pmatrix} 2 & -1 \\ 0 & 1 \end{pmatrix}$ and this has determinant $2 \neq 0$. Hence A has rank 2 and the two rows are linearly independent. Since the first two columns are linearly independent we can take the remaining two variables, z and w, as the independent variables. We have $y = w$, $2x = y - z = w - z$ and so $\left\{ \left((w - z)/2, w, z, w \right) : z \in \mathbb{R}, \, w \in \mathbb{R} \right\}$ is the solution set for this system of equations. We can write this in the form

$$\left\{ \left(\phi(z, w), z, w \right) : (z, w) \in \mathbb{R}^2 \right\}$$

where $\phi(z, w) = \left((w - z)/2, w \right)$ and in this format the solution space is the *graph* of the function ϕ (see Example 1.4).

Note that columns 1 and 3 are not linearly independent, since the corresponding 2×2 matrix $\begin{pmatrix} 2 & 1 \\ 0 & 0 \end{pmatrix}$ has zero determinant, and we cannot choose y and w (the remaining variables) as the independent variables.

Assuming that the rows of $F'(P)$ are linearly independent is equivalent to requiring that $\{ \nabla f_1(P), \ldots, \nabla f_m(P) \}$ are linearly independent vectors in $\mathbb{R}^n$. The implicit function theorem says that with this condition we can solve the *non-linear* system of Equations (2.1) near P and use the same method to identify a set of independent variables. The hypothesis of a *good* linear approximation in the definition of differentiable functions implies that the systems of equations, (2.1) and (2.2), are very close to one another.

We now state without proof the implicit function theorem.

Theorem 2.1 (Implicit Function Theorem)

Let $F \colon U \subset \mathbb{R}^n \to \mathbb{R}^m$, $m \leq n$, denote a differentiable function, let $P \in U$ and suppose $F(P) = C$ and rank $\left(F'(P) \right) = m$ (for convenience we suppose that the final m columns of $F'(P)$ are linearly independent). If $P = (p_1, \ldots, p_n)$ let $P_1 = (p_1, \ldots, p_{n-m})$ and $P_2 = (p_{n-m+1}, \ldots, p_n)$ so that $P = (P_1, P_2)$. Then there exists an open set V in $\mathbb{R}^{n-m}$ containing P_1, a differentiable function $\phi \colon V \to \mathbb{R}^m$, an open subset W of U containing P such that $\phi(P_1) = P_2$ and

$$F^{-1}(C) \cap W = \left\{ \left(X, \phi(X) \right) : X \in V \right\} = \text{graph } (\phi).$$

Thus locally every level set is a graph (Figure 2.1).

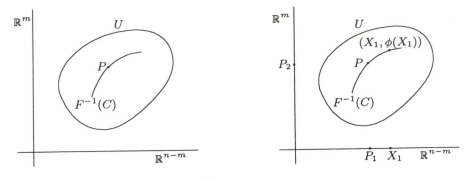

Figure 2.1.

Example 2.1

Let $F(x_1, x_2, x_3, x_4) = (x_1 x_2, x_3 x_4^2)$. We have

$$F'(x_1, x_2, x_3, x_4) = \begin{pmatrix} \nabla f_1(x_1, x_2, x_3, x_4) \\ \nabla f_2(x_1, x_2, x_3, x_4) \end{pmatrix} = \begin{pmatrix} x_2 & x_1 & 0 & 0 \\ 0 & 0 & x_4^2 & 2x_3 x_4 \end{pmatrix}$$

where $f_1(x_1, x_2, x_3, x_4) = x_1 x_2$ and $f_2(x_1, x_2, x_3, x_4) = x_3 x_4^2$. Consider the level set $F^{-1}(2, -4)$ and the point $P = (1, 2, -1, 2) \in F^{-1}(2, -4)$. We have

$$F'(1, 2, -1, 2) = \begin{pmatrix} 2 & 1 & 0 & 0 \\ 0 & 0 & 4 & -4 \end{pmatrix}.$$

If $\alpha(2, 1, 0, 0) + \beta(0, 0, 4, -1) = (0, 0, 0, 0)$ then $(2\alpha, \alpha, 4\beta, -\beta) = (0, 0, 0, 0)$ and hence $\alpha = \beta = 0$. This implies that the rows of $F'(1, 2, -1, 2)$ are linearly independent and $F'(P)$ has rank 2. This can be seen even more rapidly by finding a 2×2 submatrix with non-zero determinant, e.g. if we use columns 2 and 3 we get the matrix $\begin{pmatrix} 1 & 0 \\ 0 & 4 \end{pmatrix}$ with determinant $4 \neq 0$. It is easily checked that the following pairs of columns are linearly independent $(1, 3)$, $(1, 4)$, $(2, 3)$ and $(2, 4)$ while the pairs $(1, 2)$ and $(3, 4)$ are not linearly independent. Since columns 1 and 4 are linearly independent we know that the variables x_2 and x_3 can be chosen as a complete set of independent variables. Thus *we know* that x_1 and x_4 can be expressed as functions of x_2 and x_3 near the point $(1, 2, -1, 2)$. The implicit function theorem is *important* because it tells us that certain functions *exist* even though it does not show how to find them. In general, we would have to solve the system of Equations (2.1) to find these functions and this may often be extremely difficult or even impossible in any reasonable fashion. In our rather simple situation we have the two equations

$$\begin{aligned} x_1 x_2 &= 2 \\ x_3 x_4^2 &= -4 \end{aligned}$$

and can find a solution. We have $x_1 = 2/x_2$ and since (x_1, x_2, x_3, x_4) is close to $(1, 2, -1, 2)$ we have x_2 close to 2 and the natural domain for x_2 from this equation is $x_2 > 0$. We have $x_4^2 = -4/x_3$ and since x_3 is close to -1 we take $x_3 < 0$. Hence $-4/x_3$ is positive and $x_4 = \pm\sqrt{-4/x_3}$. Since x_4 is close to 2 we take the positive square root. Thus the function ϕ, whose *existence is foretold* by the implicit function theorem, has the form

$$\phi(x_2, x_3) = \left(\frac{2}{x_2}, +\sqrt{\frac{-4}{x_3}} \right)$$

on the open set $V = \{(x_2, x_3) : x_2 > 0, \ x_3 < 0\}$. After a rearrangement of the variables we get

$$\text{graph } (\phi) = \left\{ \left(\frac{2}{x_2}, x_2, x_3, +\sqrt{\frac{-4}{x_3}} \right) : x_2 > 0, \ x_3 < 0 \right\}$$

and, as expected, we have

$$
\begin{aligned}
F\left(\frac{2}{x_2}, x_2, x_3, +\sqrt{\frac{-4}{x_3}} \right) &= \left(\frac{2}{x_2} \cdot x_2, x_3 \left(+\sqrt{\frac{-4}{x_3}} \right)^2 \right) \\
&= \left(2, \frac{x_3(-4)}{x_3} \right) = (2, -4).
\end{aligned}
$$

An examination of the equations $x_1 x_2 = 2$ and $x_3 x_4^2 = -4$ shows that it is not possible to find, say x_3, as a function of x_1 and x_2 and thus we cannot, as expected, use x_1 and x_2 as the independent variables.

Example 2.2

Given the equations

$$
\begin{aligned}
x^2 - y^2 + u^2 + 2v^2 &= 1 & (2.3) \\
x^2 + y^2 - u^2 - v^2 &= 2 & (2.4)
\end{aligned}
$$

we wish to find all (x, y, u, v) such that near (x, y, u, v), u and v can be expressed as differentiable functions of x and y and we also wish to find $\dfrac{\partial u}{\partial x}$ and $\dfrac{\partial v}{\partial x}$ in terms of x, y, u and v.

Let $F(x, y, u, v) = (x^2 - y^2 + u^2 + 2v^2, x^2 + y^2 - u^2 - v^2)$. Then (x, y, u, v) satisfies Equations (2.3) and (2.4) if and only if $F(x, y, u, v) = (1, 2)$, i.e. if and only if $(x, y, u, v) \in F^{-1}((1, 2))$. We have

$$
F'(x, y, u, v) = \begin{pmatrix} 2x & -2y & 2u & 4v \\ 2x & 2y & -2u & -2v \end{pmatrix}.
$$

We require x and y as a complete set of independent variables and so must have linear independence of the third and fourth columns. Hence we wish to find the points (x, y, u, v) such that $\det \begin{pmatrix} 2u & 4v \\ -2u & -2v \end{pmatrix} = 4uv \neq 0$. This implies that any point (x, y, u, v) satisfying (2.3) and (2.4) with u and v both non-zero will be suitable. To compute $\dfrac{\partial u}{\partial x}$ and $\dfrac{\partial v}{\partial y}$ we could apply the chain rule to the equation

$$F\big(x, y, u(x, y), v(x, y)\big) = (1, 2)$$

or just use Equations (2.3) and (2.4) which now read

$$
\begin{aligned}
x^2 - y^2 + u(x, y)^2 + 2v(x, y)^2 &= 1 \\
x^2 + y^2 - u(x, y)^2 - v(x, y)^2 &= 2.
\end{aligned}
$$

On differentiating we get

$$2x + 2u(x, y)\frac{\partial u}{\partial x} + 4v(x, y)\frac{\partial v}{\partial x} = 0 \tag{2.5}$$

$$2x - 2u(x, y)\frac{\partial u}{\partial x} - 2v(x, y)\frac{\partial v}{\partial x} = 0. \tag{2.6}$$

The method of differentiation used to obtain Equations (2.5) and (2.6) is often called *"implicit differentiation"* especially in one-variable calculus where level sets of the form $f(x, y) = 0$ are considered.

These are just two linear equations in $\dfrac{\partial u}{\partial x}$ and $\dfrac{\partial v}{\partial x}$ which are easily solved to give

$$\frac{\partial u}{\partial x} = \frac{3x}{u} \quad \text{and} \quad \frac{\partial v}{\partial x} = \frac{-2x}{v}.$$

Notice how we need u and v to be both non-zero. In most cases of this type it is not possible to find *explicit* formulae for u and v by solving equations similar to (2.3) and (2.4) – hence the *implicit* in the implicit function theorem. So although in general we cannot find explicit formulae for the dependent variables in terms of the independent variables we can find the partial derivatives in terms of the independent and dependent variables. We choose this particular example because we are able to verify our formulae for $\dfrac{\partial u}{\partial x}$ and $\dfrac{\partial v}{\partial x}$. Adding (2.3) and (2.4) gives $2x^2 + v^2 = 3$ and hence $v = \pm(3 - 2x^2)^{1/2}$ where we take the appropriate sign depending on the value of v. We have

$$\frac{\partial v}{\partial x} = \pm\frac{1}{2}(3 - 2x^2)^{-1/2}(-4x) = \frac{-2x}{\pm(3 - 2x^2)^{1/2}} = \frac{-2x}{v}.$$

Subtracting (2.1) and (2.2) gives

$$-2y^2 + 2u^2 + 3v^2 = -1 = -2y^2 + 2u^2 + 3(3 - 2x^2)$$

i.e. $2u^2 = 2y^2 + 6x^2 - 10$ and $u = \pm(3x^2 + y^2 - 5)^{1/2}$. Hence

$$\frac{\partial u}{\partial x} = \pm\frac{1}{2}(3x^2 + y^2 - 5)^{-1/2} \cdot 6x = \frac{3x}{\pm(3x^2 + y^2 - 5)^{1/2}} = \frac{3x}{u}$$

and this agrees with our earlier calculation.

We now return to the general situation and let

$$F = (f_1, \ldots, f_m) : U \text{ (open)} \subset \mathbb{R}^n \to \mathbb{R}^m$$

denote a differentiable function which has full rank at the point P in U. Let $F(P) = C$. By the definition of differentiability the function $H : \mathbb{R}^n \to \mathbb{R}^m$ defined by

$$H(X) = F(P) + F'(P) \circ (X - P)$$

is the closest linear approximation to F near P. Since $H(P) = F(P) = C$ the level set $H^{-1}(C)$ also passes through the point P and we call this set the *tangent space* to $F^{-1}(C)$ at P. We have

$$H^{-1}(C) = \left\{X \in \mathbb{R}^n : F'(P) \circ (X - P) = 0\right\} = P + \left\{X \in \mathbb{R}^n : F'(P) \circ X = 0\right\}.$$

Since

$$F'(P) = \left(\frac{\partial f_i}{\partial x_j}(P)\right)_{\substack{1 \le i \le m \\ 1 \le j \le n}}$$

the set of all X satisfying the equation $F'(P) \circ X = 0$ is precisely the set of all solutions $\{x_1, \ldots, x_n\}$ of the system of homogeneous linear Equations (2.2) that we encountered earlier. Since F has full rank at P this space of solutions forms an $(n - m)$–dimensional subspace of $\mathbb{R}^n$ and, moreover, as an inspection of (2.2) immediately shows we have

$$\left\{X \in \mathbb{R}^n : F'(P) \circ X = 0\right\} = \left\{X \in \mathbb{R}^n : \langle X, \nabla f_i(P)\rangle = 0, \, i = 1, \ldots, m\right\}.$$

Thus the tangent space consists of the vectors which are perpendicular to the gradients of the component functions *transferred* to the point P (see Figure 2.2). The subspace of $\mathbb{R}^n$ spanned by the gradients, i.e. the set

$$\left\{\sum_{i=1}^m \lambda_i \nabla f_i(P) : \lambda_i \in \mathbb{R}\right\}$$

is m–dimensional. We define the *normal space* to the level set at P as the set

$$P + \left\{\sum_{i=1}^m \lambda_i \nabla f_i(P) : \lambda_i \in \mathbb{R}\right\}.$$

If the tangent space or normal space is two-dimensional we use the term *tangent plane* and *normal plane* respectively and if it is one-dimensional we use *tangent line* and *normal line* respectively. The tangent space and normal space are both translates of vector subspaces of $\mathbb{R}^n$ to the point P. The tangent space gives us the subspace which fits closest to the level set of F at P while the normal subspace gives us the directions which are – roughly speaking – perpendicular to the surface near P.

In $\mathbb{R}^n$ there are various ways of presenting lines, planes, etc. The *normal form* consists of a description as the set of points satisfying a set of equations while the *parametric form* is in terms of independent variables and this, as we shall see in Chapters 7 and 10, is almost a parametrization of the space.

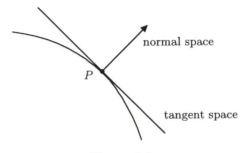

Figure 2.2.

Example 2.3

Let S denote the set of all points in $\mathbb{R}^3$ which satisfy the equation $x^2 + 2y^2 - 5z^2 = 1$. We wish to find the tangent space and the normal line at the point $(2, -1, 1)$ on S. Let $f(x, y, z) = x^2 + 2y^2 - 5z^2$. Then $S = f^{-1}(1)$. The tangent plane at (x_0, y_0, z_0) in normal form is

$$\Big\{ (x, y, z) : (x - x_0, y - y_0, z - z_0) \cdot \nabla f(x_0, y_0, z_0) = 0 \Big\}$$

$$= \Big\{ (x, y, z) : (x - x_0) \frac{\partial f}{\partial x}(x_0, y_0, z_0) + (y - y_0) \frac{\partial f}{\partial y}(x_0, y_0, z_0)$$

$$+ (z - z_0) \frac{\partial f}{\partial z}(x_0, y_0, z_0) = 0 \Big\}$$

Now $\nabla f(x, y, z) = (2x, 4y, -10z)$ and $\nabla f(2, -1, 1) = (4, -4, -10)$ and so the tangent plane at $(2, -1, 1)$ is

$$\Big\{ (x, y, z) : (x - 2)4 + (y + 1)(-4) + (z - 1)(-10) = 0 \Big\}$$

$$= \Big\{ (x, y, z) : 2x - 2y - 5z = 1 \Big\}.$$

The normal line at (x_0, y_0, z_0) is the line through (x_0, y_0, z_0) in the direction of $\nabla f(x_0, y_0, z_0)$. In our case we have, in parametric form, the normal line

$$\left\{ (2, -1, 1) + t(4, -4, -10) : t \in \mathbb{R} \right\} = \left\{ (2 + 4t, -1 - 4t, 1 - 10t) : t \in \mathbb{R} \right\}.$$

To change this into normal form let $x = 2 + 4t$, $y = -1 - 4t$ and $z = 1 - 10t$. Solving for t we get

$$\frac{x - 2}{4} = \frac{y + 1}{-4} = \frac{z - 1}{-10} = t$$

and obtain the normal form

$$\left\{ (x, y, z) : \frac{x - 2}{4} = \frac{y + 1}{-4} = \frac{z - 1}{-10} \right\}.$$

To find the tangent plane in parametric form we must find two linearly independent vectors which are perpendicular to $\nabla f(x_0, y_0, z_0)$. Applying the observation that (x, y) and $(-y, x)$ are perpendicular vectors in $\mathbb{R}^2$ to different pairs of coordinates in $\mathbb{R}^3$ we see easily that

$$(4, -4, -10) \cdot (4, 4, 0) = 0$$

and

$$(4, -4, -10) \cdot (10, 0, 4) = 0$$

and, moreover, $(4, 4, 0)$ and $(10, 0, 4)$ are linearly independent. In parametric form the tangent space at $(2, -1, 1)$ is

$$(2, -1, 1) + \left\{ x(4, 4, 0) + y(10, 0, 4) : x, y \in \mathbb{R} \right\}$$
$$= \left\{ (2 + 4x + 10y, -1 + 4x, 1 + 4y) : x, y \in \mathbb{R} \right\}.$$

Example 2.4

We wish to find the normal plane and the tangent line to the set of points satisfying

$$x^2 + y^2 - 2z^2 = 2 \quad \text{and} \quad xyz = 2$$

at the point $(\sqrt{2}, \sqrt{2}, 1)$.

Let $F(x, y, z) = (x^2 + y^2 - 2z^2, xyz)$. Then the set of points which satisfy the above equations form the level set $F^{-1}(2, 2)$. We have

$$F'(\sqrt{2}, \sqrt{2}, 1) = \begin{pmatrix} 2\sqrt{2} & 2\sqrt{2} & -4 \\ \sqrt{2} & \sqrt{2} & 2 \end{pmatrix}.$$

The final two columns form a 2×2 matrix with non-zero determinant and hence F has full rank at $(\sqrt{2}, \sqrt{2}, 1)$. The normal plane at $(\sqrt{2}, \sqrt{2}, 1)$ has parametric form

$$
(\sqrt{2}, \sqrt{2}, 1) + \left\{ x(2\sqrt{2}, 2\sqrt{2}, -4) + y(\sqrt{2}, \sqrt{2}, 2) : x, y \in \mathbb{R} \right\}
$$
$$
= \left\{ (\sqrt{2} + 2\sqrt{2}x + \sqrt{2}y, \sqrt{2} + 2\sqrt{2}x + \sqrt{2}y, 1 - 4x + 2y) : x, y \in \mathbb{R} \right\}.
$$

To find the normal plane in normal form we must find a non-zero vector perpendicular to both $(2\sqrt{2}, 2\sqrt{2}, -4)$ and $(\sqrt{2}, \sqrt{2}, 2)$. The cross product (see Chapters 6 and 7) of the two given vectors is of the required type but we take a first-principles approach here. This amounts to finding (a, b, c) such that $(\sqrt{2}, \sqrt{2}, 2) \cdot (a, b, c) = 0$ and $(2\sqrt{2}, 2\sqrt{2}, -4) \cdot (a, b, c) = 0$. We thus have to solve the system of equations

$$
\begin{aligned}
2\sqrt{2}a + \sqrt{2}b + 2c &= 0 \\
\sqrt{2}a + \sqrt{2}b - c &= 0.
\end{aligned}
$$

Subtracting we see that $c = 0$ and we can take $a = -b = 1$. Hence $(1, -1, 0)$ is a suitable vector. The normal plane (in normal form) is

$$
\left\{ (x, y, z) : (x - \sqrt{2}, y - \sqrt{2}, z - 1) \cdot (1, -1, 0) = 0 \right\}
$$
$$
= \left\{ (x, y, z) : x - \sqrt{2} - y + \sqrt{2} = 0 \right\}
$$
$$
= \left\{ (x, y, z) : x - y = 0 \right\}.
$$

The tangent line in normal form is

$$
(\sqrt{2}, \sqrt{2}, 1) + \left\{ (x, y, z) : 2\sqrt{2}x + 2\sqrt{2}y - 4z = 0, \sqrt{2}x + \sqrt{2}y + 2z = 0 \right\}
$$
$$
= (\sqrt{2}, \sqrt{2}, 1) + \left\{ (x, y, z) : y = -x, z = 0 \right\}
$$
$$
= \left\{ (x, y, z) : x + y = 2\sqrt{2}, z = 1 \right\}.
$$

From the second of these equations we see that the tangent line in parametric form is

$$
\left\{ (\sqrt{2} + t, \sqrt{2} - t, 1) : t \in \mathbb{R} \right\}.
$$

EXERCISES

2.1 Let $F_i \colon \mathbb{R}^4 \to \mathbb{R}^i$

$$
\begin{aligned}
F_1(x_1, x_2, x_3, x_4) &= x_1^2 - x_2^2, \quad P_1 = (1, 2, 0, -1) \\
F_2(x_1, x_2, x_3, x_4) &= (x_1^2 - x_2^2, x_3^2 - x_4^2), \quad P_2 = (1, 0, 2, -1) \\
F_3(x_1, x_2, x_3, x_4) &= (x_1^2 - x_2^2, x_3^2 - x_4^2, x_4^2 - x_1^2), \quad P_3 = (1, 2, 3, 4).
\end{aligned}
$$

Calculate $F_i'(X)$ for $X \in \mathbb{R}^4$ and find all X such that F_i has full rank at X. When F_i has full rank find all subsets of $\{x_1, x_2, x_3, x_4\}$ which can be taken as complete sets of independent variables. If $F_i(P_i) = C_i$ and F_i has full rank at P_i find a function $\phi_i \colon \mathbb{R}^{4-i} \to \mathbb{R}^i$ such that $F_i^{-1}(C_i) = \mathrm{graph}(\phi_i)$ near P_i.

2.2 If $u(x, y)$ and $v(x, y)$ are defined by the equations $u \cos v = x$ and $u \sin v = y$ find $\dfrac{\partial u}{\partial x}$ and $\dfrac{\partial v}{\partial x}$ by

(i) finding explicit formulae for u and v

(ii) using implicit differentiation.

2.3 Let $F(x_1, x_2, x_3, x_4) = (x_1^2 x_2^2, x_1 x_2 x_3, x_4^2)$. Find $F'(1, 2, 3, 4)$. Let $A = F'(1, 2, 3, 4)$. Display the system of equations $AX = 0$. Solve this system of equations and find a basis for the space of solutions. Using your set of solutions find the tangent space to the level set of F at $(1, 2, 3, 4)$.

2.4 (a) Find in normal and parametric form the normal line and the tangent plane to the surface $z = xe^y$ at the point $(1, 0, 1)$.

(b) The surfaces $x^2 + y^2 - z^2 = 1$ and $x + y + z = 5$ intersect in a curve Γ. Find the equation in parametric form of the tangent line to Γ at the point $(1, 2, 2)$.

2.5 Find the equation of the plane passing through the points $(1, 2, 3)$ and $(4, 5, 6)$ which is perpendicular to the plane $7x + 8y + 9z = 10$.

2.6 Find the equation of the tangent plane to $\sqrt{x} + \sqrt{y} + \sqrt{z} = 4$ at the point $(1, 4, 1)$.

2.7 Find the tangent planes at $(1/\sqrt{2}, 1/4, 1/4)$ and $(\sqrt{3}/2, 0, 1/4)$ to the ellipsoid $x^2 + 4y^2 + 4z^2 = 1$. Find the line of intersection of these two planes. Show that this line is tangent to the sphere $x^2 + y^2 + z^2 = k$ for exactly one value of k and find this value.

2.8 Find the coordinates of the four points where the hyperbola $x^2 - y^2 = 1$ and the ellipse $x^2 + 2y^2 = 4$ intersect. If (a, b), $a > b > 0$, is one of these points show that the tangent line to the hyperbola at (a, b) coincides with the normal line to the ellipse at (a, b). Show that the tangent lines to the hyperbola at the four points enclose a parallelogram and find its area.

2.9 Find the direction of the normal line at the point $(1, 1, 4)$ to the paraboloid $z = x^2 + y^2 + 2$. Find the tangent plane in normal form at this point. Show that the normal line meets the paraboloid again at the point $(-5/4, -5/4, 41/8)$. If θ is the angle between the normal line through the point $(1, 1, 4)$ and the normal line through the point $(-5/4, -5/4, 41/8)$ show that $\sin \theta = 1/\sqrt{3}$.

2.10 Consider the subset S of $\mathbb{R}^3$ which lies above the (x, y)–plane and which is characterised by the property:

$$p \in S \iff \qquad \text{the distance from } p \text{ to the } xy\text{-plane is}$$
$$\text{the logarithm of its distance to the } z\text{-axis.}$$

Describe S as a level set and as a graph. Find the normal line and the tangent plane to S at the point $(1, -1, \log 2/2)$.

2.11 Let $S_1 = \{(x, y, z) \in \mathbb{R}^3 : y = f(x)\}$ denote a cylinder and let S_2 denote the level set $z^2 + 2zx + y = 0$. If S_1 is tangent to S_2 at all points of contact find f.

2.12 Let V denote a proper subspace of $\mathbb{R}^n$, i.e. $0 \neq V \neq \mathbb{R}^n$. Let $\{\mathbf{b}_1, \ldots, \mathbf{b}_n\}$ denote an orthonormal basis for $\mathbb{R}^n$ and let $\{\mathbf{b}_1^*, \ldots, \mathbf{b}_n^*\}$ denote the corresponding dual basis for $\mathbb{R}_n^*$. If $\{\mathbf{b}_1, \ldots, \mathbf{b}_m\}$ is a basis for V use the dual basis to describe V as a level set. Show that V, when identified with a level set, is its own tangent space at each point. Find the normal space at each point.

2.13 Let $\mathbf{a} = (a_1, a_2)$ and $\mathbf{b} = (b_1, b_2)$ denote two non-zero vectors in $\mathbb{R}^2$ making angles θ_1 and θ_2, respectively, with the positive x-axis. Show that $\cos \theta_1 = a_1/\|\mathbf{a}\|$ and $\sin \theta_1 = a_2/\|\mathbf{a}\|$. Show that $\theta_2 - \theta_1$ is the angle between $\mathbf{a}$ and $\mathbf{b}$ and, by expanding $\cos(\theta_2 - \theta_1)$, prove that $\cos(\theta_2 - \theta_1) = \mathbf{a} \cdot \mathbf{b}/\|\mathbf{a}\| \cdot \|\mathbf{b}\|$. Prove the same result for arbitrary vectors in $\mathbb{R}^n$.

3

Lagrange Multipliers

Summary. *We develop the method of Lagrange multipliers to find the maximum and minimum of a function with constraints.*

We consider the problem of finding the maximum and minimum of a sufficiently regular function g of n variables subject to the constraints

$$\left.\begin{array}{ll} f_1(x_1,\ldots,x_n) & = c_1 \\ \qquad\vdots \\ f_m(x_1,\ldots,x_n) & = c_m. \end{array}\right\} \tag{3.1}$$

To apply differential calculus we suppose there exists an open subset U of $\mathbb{R}^n$ such that each f_i is defined and differentiable on U and that all points which satisfy (3.1), that we wish to consider, lie in U. We also require g to be smooth in some fashion and it is convenient to suppose now that g is also differentiable on U.

If $F = (f_1,\ldots,f_m)$ then $F\colon U \subset \mathbb{R}^n \to \mathbb{R}^m$ is differentiable and our problem can be restated as that of finding the maximum and minimum of g on the level set $F^{-1}(C)$ where $C = (c_1,\ldots,c_m)$. To apply the methods of the previous chapter we suppose that F has full rank on $F^{-1}(C)$ or equivalently that $\{\nabla f_1(X),\ldots,\nabla f_m(X)\}$ are linearly independent vectors for all X in $F^{-1}(C)$.

Suppose g has a local maximum (or minimum) on $F^{-1}(C)$ at the point P. Since we only need to examine g near P we may suppose that $F^{-1}(C)$ is the graph of a function ϕ of $n-m$ variables. By rearranging the variables, if necessary, we can assume that $\phi\colon V \subset \mathbb{R}^{n-m} \to \mathbb{R}^m$, V open in $\mathbb{R}^{n-m}$, that P_1 (the first $n-m$ coordinates of P) lies in V and $P = (P_1, \phi(P_1))$.

23

For the sake of clarity let

$$X_1 = (x_1, \ldots, x_{n-m}) \in \mathbb{R}^{n-m}, \quad X_2 = (x_{n-m+1}, \ldots, x_n) \in \mathbb{R}^m$$

and let $\phi = (\phi_{n-m+1}, \ldots, \phi_n)$ where each ϕ_i is a real-valued function on V. Let $O_{p,q}$ denote the $p \times q$ matrix with all entries zero and let A be the $n \times (n-m)$ matrix

$$\begin{pmatrix} 1 & \cdots & 0 \\ \vdots & \ddots & \\ 0 & \cdots & 1 \\ & \nabla\phi_{n-m+1}(P_1) & \\ & \vdots & \\ & \nabla\phi_n(P_1) & \end{pmatrix} .$$

With this notation $X = (X_1, X_2) = (x_1, \ldots, x_{n-m}, x_{n-m+1}, \ldots, x_n) \in \mathbb{R}^n$, $F(X_1, \phi(X_1)) = 0$ for all $X_1 \in V$, the function G defined by $G(X_1) = g(X_1, \phi(X_1))$ has a local maximum (or minimum) on V at P_1 and A is the derivative at P_1 of the mapping

$$X_1 \in V \subset \mathbb{R}^{n-m} \longrightarrow (X_1, \phi(X_1)) \in \mathbb{R}^n .$$

By restricting G to the one-dimensional sections (i.e. lines) passing through P_1 and applying one-variable calculus we see that $D_\mathbf{v}G(P_1) = 0$ for all $\mathbf{v} \in \mathbb{R}^{n-m}$. In particular, all first-order partial derivatives of G at P_1 vanish and P_1 is a *critical point* of G, i.e. $\nabla G(P_1) = 0$ (in $\mathbb{R}^{n-m}$). Experience suggests that we examine the consequence of this fact. To calculate $\nabla G(P_1)$ we apply the *chain rule* to the decomposition

$$X_1 \longrightarrow (X_1, \phi(X_1))$$
$$\|$$
$$(X_1, X_2) \longrightarrow g(X_1, X_2)$$

of G and obtain the matrix equation

$$\nabla g(P) \circ A = O_{1,n-m} . \tag{3.2}$$

Applying the chain rule to the equation $F(X, \phi(X)) = 0$ we obtain

$$F'(P) \circ A = O_{m,n-m}$$

and taking the ith row of this equation, $1 \leq i \leq m$, we have

$$\nabla f_i(P) \circ A = O_{1,n-m} . \tag{3.3}$$

Equations (3.2) and (3.3) tell us that the $m + 1$ vectors $\{\nabla g(P), \nabla f_1(P), \ldots,$ $\nabla f_m(P)\}$ are solutions of the system of $(n - m)$ homogeneous linear equations in n unknowns,

$$Y \circ A = O_{1,n-m} \tag{3.4}$$

with matrix of coefficients A and unknown $Y = (y_1, \ldots, y_n) \in \mathbb{R}^n$. Since the first m rows of A form the identity $m \times m$ matrix they are linearly independent and hence the space of solutions for the system (3.4) is $n - (n - m) = m$-dimensional. By our hypothesis $\{\nabla f_1(P), \ldots, \nabla f_m(P)\}$ is a linearly independent set of vectors and hence forms a basis for the space of solutions. Since $\nabla g(P)$ is also a solution for (3.4) there exist m scalars, called *Lagrange multipliers*, $\lambda_1, \ldots, \lambda_m$, such that

$$\nabla g(P) = \lambda_1 \nabla f_1(P) + \cdots + \lambda_m \nabla f_m(P) . \tag{3.5}$$

Equation (3.5), in terms of coordinates, is a set of m equations, and these together with the system of Equations (3.1) gives $n + m$ equations in the $n + m$ unknowns $\{x_1, \ldots, x_n, \lambda_1, \ldots, \lambda_m\}$. The *method of Lagrange multipliers* consists in solving these equations. The solutions are the critical points of g on $F^{-1}(C)$ and contain the local maxima and minima of g. To determine if g has a maximum and minimum on $F^{-1}(C)$ requires further on-the-spot investigation as we shall see in examples.

Frequently we are interested in finding the maximum and minimum on $\overline{U} \cap F^{-1}(C)$ where U is open in $\mathbb{R}^n$ and $\overline{U}$ is the closure of U in $\mathbb{R}^n$. The set $\overline{U}$ consists of all points which can be reached from U, i.e.

$$\overline{U} = \{X \in \mathbb{R}^n; \text{ there exists } (X_n)_n \in U \text{ with } \|X_n - X\| \to 0 \text{ as } n \to \infty\} .$$

In this case we apply the method of Lagrange multipliers to $U \cap F^{-1}(C)$ and examine separately the values of g on $(\overline{U} \setminus U) \cap F^{-1}(C)$ – these points are often easily identified since they lie on the *boundary* of U.

Example 3.1

To find the maximum and minimum of $x + y + z$ subject to the constraints $x^2 + y^2 = 2$ and $x + z = 1$. Let $F = (f_1, f_2)$ where $f_1(x, y, z) = x^2 + y^2$ and $f_2(x, y, z) = x + z$ and let $g(x, y, z) = x + y + z$. We wish to find the maximum and minimum of g on the set $F^{-1}(2, 1)$. We have

$$F'(x, y, z) = \begin{pmatrix} 2x & 2y & 0 \\ 1 & 0 & 1 \end{pmatrix} .$$

On $F^{-1}(2, 1)$ we have $x^2 + y^2 = 2$ and so x and y cannot both be zero at the same time. Thus the rows of $F'(x, y, z)$ are linearly independent when

$(x, y, z) \in F^{-1}(2, 1)$. Hence ∇f_1 and ∇f_2 are linearly independent on $F^{-1}(2, 1)$ and we may apply the method of Lagrange multipliers. The level set

$$F^{-1}(2, 1) = \{(x, y, z) : x^2 + y^2 = 2\} \cap \{(x, y, z) : x + z = 1\}$$

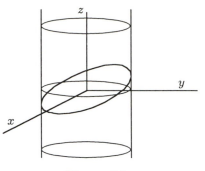

Figure 3.1.

is the intersection of a cylinder parallel to the z-axis and a plane (why?) (see Figure 3.1). This gives us a closed ellipse and compact set in $\mathbb{R}^3$. By the fundamental existence theorem g has a maximum and a minimum on $F^{-1}(2, 1)$. This apparently rather theoretical information has practical implications since we now know, before doing *any* calculations, that the method of Lagrange multipliers will yield *at least* two solutions and the maximum (minimum) solution will be the maximum (minimum) of g on $F^{-1}(2, 1)$. By the method of Lagrange multipliers there exist, at local maxima and minima, scalars λ_1 and λ_2 such that

$$\nabla g(x, y, z) = \lambda_1 \nabla f_1(x, y, z) + \lambda_2 \nabla f_2(x, y, z)$$

i.e.

$$(1, 1, 1) = \lambda_1(2x, 2y, 0) + \lambda_2(1, 0, 1) .$$

In terms of coordinates this is equivalent to the following system of equations

$$\left. \begin{array}{l} 1 = 2\lambda_1 x + \lambda_2 \\ 1 = 2\lambda_1 y \\ 1 = \lambda_2 \end{array} \right\} \implies 1 = 2\lambda_1 x + 1 \implies \lambda_1 x = 0.$$

Hence $\lambda_1 = 0$ or $x = 0$. However, the second equation implies $1 = 2\lambda_1 y$ so $\lambda_1 \neq 0$. If $x = 0$ then $f_1(x, y, z) = y^2 = 2$ and $y = \pm\sqrt{2}$, while $f_2(x, y, z) = x + z = 1$ implies $z = 1$. Our only solutions are $(0, \sqrt{2}, 1)$ and $(0, -\sqrt{2}, 1)$. Since $g(0, \sqrt{2}, 1) = \sqrt{2} + 1$ and $g(0, -\sqrt{2}, 1) = -\sqrt{2} + 1$ it follows that $\sqrt{2} + 1$ and $-\sqrt{2} + 1$ are the maximum and minimum of g on $F^{-1}(2, 1)$.

Of course, substituting the constraint $x + z = 1$ into $g = 0$ implies $g(x, y, z) = g(x, y, 1 - x) = 1 + y$ and our problem reduces to finding the maximum and minimum of $1 + y$ on the set $x^2 + y^2 = 2$. On this set $-\sqrt{2} \le y \le \sqrt{2}$ and hence the maximum is $1 + \sqrt{2}$ and the minimum $1 - \sqrt{2}$. This verifies the solution obtained using Lagrange multipliers and also reminds us that it is always worthwhile thinking about a problem before attempting any solution.

Example 3.2

The *geometric mean* of n positive numbers $x_1, \ldots, x_n$, $(x_1 \cdots x_n)^{1/n}$ is always less than or equal to the *arithmetic mean* $\dfrac{x_1 + x_2 + \cdots + x_n}{n}$.

In this example (see also Example 3.4) we standardise one of the quantities and generate a Lagrange multiplier type problem. Let $g(x_1, \ldots, x_n) = x_1 \cdots x_n$ and $f(x_1, \ldots, x_n) = x_1 + x_2 + \cdots + x_n$. We begin by finding the maximum of g on the set

$$f^{-1}(1) \cap \left\{ (x_1, \ldots, x_n) : x_i \ge 0 \text{ for all } i \right\}.$$

It is easily checked that this is a compact set and, by the fundamental existence theorem, the maximum exists. On the set $f^{-1}(1) \cap \left\{ (x_1, \ldots, x_n) : x_i > 0 \right\}$ we may apply the method of Lagrange multipliers. We have

$$
\begin{aligned}
\nabla g(x_1, \ldots, x_n) &= (x_2 \cdots x_n, x_1 x_3 \cdots x_n, \ldots, x_1 x_2 \cdots x_{n-1}) \\
&= x_1 \cdots x_n \left(\tfrac{1}{x_1}, \tfrac{1}{x_2}, \ldots, \tfrac{1}{x_n} \right)
\end{aligned}
$$

and

$$\nabla f(x_1, \ldots, x_n) = (1, 1, \ldots, 1).$$

If $\nabla g(x_1, \ldots, x_n) = \lambda \nabla f(x_1, \ldots, x_n)$ then $x_1 \cdots x_n / x_i = \lambda$ and $x_i = x_1 \cdots x_n / \lambda$ for all i. This shows $x_1 = x_2 = \ldots = x_n$. Since $x_1 + x_2 + \cdots + x_n = 1$ we have $x_i = 1/n$ for all i and $g(1/n, 1/n, \ldots, 1/n) = n^{-n}$. As $g(x_1, \ldots, x_n) = 0$ whenever one of the x_i's is equal to zero it follows that the maximum of g, on the set $f(x_1, \ldots, x_n) = 1$ and $x_i \ge 0$ all i, is $(1/n)^n$.

If x_i, $i = 1, \ldots, n$, are arbitrary positive numbers let $y_i = x_i \left/ \sum_{j=1}^n x_j \right.$ for each i. We have

$$\sum_{i=1}^n y_i = \sum_{i=1}^n x_i \left/ \sum_{j=1}^n x_j \right. = 1,$$

and, by the first part,

$$\frac{\dfrac{x_1 \cdots x_n}{n}}{\left(\sum_{j=1}^n x_j \right)^n} = y_1 \cdots y_n \le \left(\frac{1}{n} \right)^n.$$

Hence

$$x_1 \cdots x_n \le \left(\frac{1}{n} \sum_{j=1}^{n} x_j \right)^n$$

and, as required,

$$(x_1 \cdots x_n)^{1/n} \le \frac{1}{n}(x_1 + x_2 + \cdots + x_n) \ .$$

Notice that although the number of variables, n, could be large symmetry led to rather simple equations. In mathematics symmetry compensates for size.

Example 3.3

To find the dimensions of the box (Figure 3.2) of maximum volume V given that the surface area S is 10m^2.

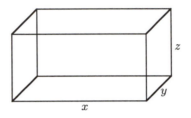

Figure 3.2.

We wish to maximise $V(x, y, z) = xyz$ subject to the constraint $S(x, y, z) = 2(xy + yz + xz) = 10$. Physical constraints imply that $x \ge 0$, $y \ge 0$ and $z \ge 0$. The set $U = \{(x, y, z) : x > 0, y > 0, z > 0\}$ is open and our problem is to determine the maximum of V on $\overline{U} \cap S$. The method of Lagrange multipliers will identify a set which contains all local maxima and minima in $U \cap S$. We proceed to do this now. We have

$$\nabla V(x, y, z) = (yz, xz, xy)$$

and

$$\nabla S(x, y, z) = \big(2(y + z), 2(x + z), 2(x + y)\big) \ .$$

Since x, y and z are all positive on U we have $\nabla V(x, y, z) \ne (0, 0, 0)$. Applying the method of Lagrange multipliers we consider

$$\nabla V = \lambda \nabla S$$

and obtain the following system of equations

$$yz = 2\lambda(y + z) \qquad\qquad xyz = 2\lambda(xy + xz) \qquad\qquad (3.6)$$
$$xz = 2\lambda(x + z) \qquad \Longrightarrow\ xyz = 2\lambda(xy + yz) \qquad\qquad (3.7)$$
$$xy = 2\lambda(x + y) \qquad\qquad xyz = 2\lambda(xz + yz) \ . \qquad\qquad (3.8)$$

Hence $\lambda \neq 0$ and, on dividing, we get

$$1 = \frac{(3.6)}{(3.7)} = \frac{(3.6)}{(3.8)} = \frac{xy + xz}{xy + yz} = \frac{xy + xz}{xz + yz} \ .$$

Cross multiplying implies

$$xy + xz = xy + yz = xz + yz \ .$$

Hence $xz = yz$ and $xy = xz$. Since $x \neq 0$, $y \neq 0$ and $z \neq 0$ this implies $x = y = z$ and $S(x, y, z) = 10$ now shows that $6x^2 = 10$. Hence $x = y = z = (5/3)^{1/2}$. Let $P = \left((5/3)^{1/2}, (5/3)^{1/2}, (5/3)^{1/2}\right)$. We have $V(P) = (5/3)^{3/2}$.

We now wish to show that V takes its *maximum* value, subject to the constraint $S = 10$, at P. The set $\overline{U} \cap S^{-1}(10)$ is closed but not bounded and so the fundamental existence theorem for maxima and minima cannot be applied directly. In such cases ad hoc methods are necessary and the following approach is sometimes useful. For $r > 0$ let

$$U_r = \left\{(x, y, z) : 0 \leq x \leq r, 0 \leq y \leq r, 0 \leq z \leq r\right\} \ .$$

The set U_r is compact (i.e. closed and bounded) and as r increases to infinity U_r expands, inside $\overline{U}$, and in the limit, i.e. at infinity, it covers $\overline{U}$. If V does have a maximum on $\overline{U} \cap S^{-1}(10)$ then it must lie in some U_r. Since $V(P) > 1$ our strategy is to show that for r sufficiently large we have $V(x, y, z) \leq 1$ whenever $(x, y, z) \notin U_r$. This will imply that the maximum of V on U_r, which exists by the fundamental existence theorem, equals the maximum of V on U. If $(x, y, z) \notin U_r$ then one of x, y, z, say x, is greater than r. If $S(x, y, z) = 10$ then $2xy \leq 10$ and $2xz \leq 10$. Hence $y \leq 10/2x = 5/x$ and $z \leq 10/2x = 5/x \leq 5/r$ and

$$V(x, y, z) = xyz \leq x \cdot \frac{5}{x} \cdot \frac{5}{r} = \frac{25}{r} \ . \qquad\qquad (3.9)$$

In particular, if $r = 25$ then $V(x, y, z) \leq 1$ for all $(x, y, z) \notin U_{25}$ satisfying $S(x, y, z) = 10$. Since $P \in U_{25}$ and $V(P) > 1$ this implies

$$\text{maximum}\left\{V(x, y, z) : (x, y, z) \in \overline{U} \cap S^{-1}(10)\right\}$$
$$= \text{maximum}\left\{V(x, y, z) : (x, y, z) \in U_{25} \cap S^{-1}(10)\right\} \ .$$

Since $U_{25} \cap S^{-1}(10)$ is compact and V is continuous the fundamental existence theorem implies that V has a maximum on $U_{25} \cap S^{-1}(10)$. The maximum occurs either inside or on the boundary. If it occurs inside then our use of Lagrange multipliers implies that it occurs at the point P. On the boundary of U_{25} we have either at least one coordinate zero, in which case $V = 0$, or at least one coordinate equal to 25 in which case (3.9) implies $V \leq 1$ and we conclude that the maximum cannot occur on the boundary. We have thus shown that the absolute maximum of V occurs at the point $P = \left((5/3)^{1/2}, (5/3)^{1/2}, (5/3)^{1/2}\right)$ and equals $(5/3)^{3/2}$.

Example 3.4

In this example we prove the *Cauchy–Schwarz inequality*

$$\left| \sum_{i=1}^{n} a_i b_i \right|^2 \leq \left(\sum_{i=1}^{n} a_i^2 \right) \left(\sum_{i=1}^{n} b_i^2 \right)$$

where a_i and b_i are arbitrary real numbers. This is one of the best known and most widely used inequalities in mathematics. If we use vector notation, i.e. let $\mathbf{a} = (a_1, \ldots, a_n)$ and $\mathbf{b} = (b_1, \ldots, b_n)$, and take square roots we can rewrite the inequality as

$$|\langle \mathbf{a}, \mathbf{b} \rangle| \leq \|\mathbf{a}\| \cdot \|\mathbf{b}\| . \tag{3.10}$$

The advantages of vector notation in this proof are obvious. Let $g : \mathbb{R}^{2n} \to \mathbb{R}$ be defined by $g(\mathbf{a}, \mathbf{b}) = \langle \mathbf{a}, \mathbf{b} \rangle$. Motivated by Example 3.2 and the form of the inequality given in (3.10) we first find the maximum and minimum of g on the set $\|\mathbf{a}\|^2 = \|\mathbf{b}\|^2 = 1$.

Let $f_1 : \mathbb{R}^{2n} \to \mathbb{R}$ and $f_2 : \mathbb{R}^{2n} \to \mathbb{R}$ be defined by

$$f_1(\mathbf{a}, \mathbf{b}) = \|\mathbf{a}\|^2 = \langle \mathbf{a}, \mathbf{a} \rangle = \sum_{i=1}^{n} a_i^2$$

and

$$f_2(\mathbf{a}, \mathbf{b}) = \|\mathbf{b}\|^2 = \langle \mathbf{b}, \mathbf{b} \rangle = \sum_{i=1}^{n} b_i^2 .$$

We wish to maximise $g(\mathbf{a}, \mathbf{b}) = \sum_{i=1}^{n} a_i b_i = \langle \mathbf{a}, \mathbf{b} \rangle$ on the set $S = \{(\mathbf{a}, \mathbf{b}) : \|\mathbf{a}\|^2 = \|\mathbf{b}\|^2 = 1\}$. The set S is easily seen to be compact and hence g has a maximum and minimum on S. We have

$$\nabla f_1(\mathbf{a}, \mathbf{b}) = 2(a_1, \ldots, a_n, 0, \ldots, 0) = 2(\mathbf{a}, \mathbf{0})$$

and

$$\nabla f_2(\mathbf{a}, \mathbf{b}) = 2(0, \ldots, 0, b_1, \ldots, b_n) = 2(\mathbf{0}, \mathbf{b})$$

where $\mathbf{0}$ denotes the origin in $\mathbb{R}^n$. An inspection of the matrix

$$\begin{pmatrix} \nabla f_1(\mathbf{a}, \mathbf{b}) \\ \nabla f_2(\mathbf{a}, \mathbf{b}) \end{pmatrix} = \begin{pmatrix} 2\mathbf{a} & \mathbf{0} \\ \mathbf{0} & 2\mathbf{b} \end{pmatrix}$$

shows immediately that $\nabla f_1(\mathbf{a}, \mathbf{b})$ and $\nabla f_2(\mathbf{a}, \mathbf{b})$ are linearly independent whenever $\mathbf{a} \neq 0$ and $\mathbf{b} \neq 0$ and implies that we may apply the method of Lagrange multipliers. From $\nabla g = \lambda_1 \nabla f_1 + \lambda_2 \nabla f_2$ and

$$\nabla g(\mathbf{a}, \mathbf{b}) = (b_1, \ldots, b_n, a_1, \ldots, a_n) = (\mathbf{b}, \mathbf{a})$$

we see that

$$(\mathbf{b}, \mathbf{a}) = \lambda_1(2\mathbf{a}, \mathbf{0}) + \lambda_2(\mathbf{0}, 2\mathbf{b}) = (2\lambda_1 \mathbf{a}, 2\lambda_2 \mathbf{b}) .$$

Hence $\mathbf{b} = 2\lambda_1 \mathbf{a}$ and since $\|\mathbf{a}\| = \|\mathbf{b}\| = 1$ this implies $2\lambda_1 = \pm 1$ and $\mathbf{b} = \pm \mathbf{a}$. If $\mathbf{b} = \mathbf{a}$ then $g(\mathbf{a}, \mathbf{a}) = \sum_{i=1}^{n} a_i^2 = 1$, while if $\mathbf{b} = -\mathbf{a}$ then $g(\mathbf{a}, -\mathbf{a}) = -\sum_{i=1}^{n} a_i^2 = -1$. Hence, if $\|\mathbf{a}\| = \|\mathbf{b}\| = 1$, then

$$-1 \leq g(\mathbf{a}, \mathbf{b}) \leq 1$$

and

$$\big| g(\mathbf{a}, \mathbf{b}) \big| \leq 1 .$$

If $\mathbf{a} \neq \mathbf{0}$ and $\mathbf{b} \neq \mathbf{0}$ then

$$\left| g\left(\frac{\mathbf{a}}{\|\mathbf{a}\|}, \frac{\mathbf{b}}{\|\mathbf{b}\|} \right) \right|^2 = \frac{\big| g(\mathbf{a}, \mathbf{b}) \big|^2}{\|\mathbf{a}\|^2 \|\mathbf{b}\|^2} \leq 1$$

and

$$\big| g(\mathbf{a}, \mathbf{b}) \big|^2 \leq \|\mathbf{a}\|^2 \cdot \|\mathbf{b}\|^2 .$$

This is trivially verified if either $\mathbf{a}$ or $\mathbf{b}$ is $\mathbf{0}$. Since $g(\mathbf{a}, \mathbf{b}) = \langle \mathbf{a}, \mathbf{b} \rangle$, we have proved the Cauchy–Schwarz inequality.

The above also shows that we have *equality* in the Cauchy–Schwarz inequality if and only if $\mathbf{a}$ and $\mathbf{b}$ are parallel vectors. In particular, for unit vectors $\mathbf{a}$ and $\mathbf{b}$, we have $\langle \mathbf{a}, \mathbf{b} \rangle = 1$ if and only if $\mathbf{a} = \mathbf{b}$ (see Example 8.4).

EXERCISES

3.1 Find the maximum and minimum of $xy + yz$ on the set of points which satisfy $x^2 + y^2 = 1$ and $yz = x$.

3.2 Find the highest and lowest points on the ellipse of intersection of the cylinder $x^2 + y^2 = 1$ and the plane $x + y + z = 1$.

3.3 Suppose a pentagon is composed of a rectangle surmounted by an isosceles triangle. If the length of the outer perimeter P is fixed find the maximum possible area.

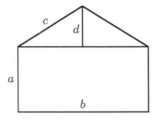

Figure 3.3.

(Hint: you must maximise $ab + \frac{1}{2}bd$ subject to the constraints $b + 2a + 2c = P$ and $(b/2)^2 + d^2 = c^2$.)

3.4 Find the minimum value of $x^2 + y^2 + z^2$ subject to the constraints $x + y - z = 0$ and $x + 3y + z = 2$. Give a *geometrical* interpretation of your answer and using your interpretation explain why $x^2 + y^2 + z^2$ has no maximum subject to these constraints.

3.5 Use the method of Lagrange multipliers with two constraints to find the minimum value of the function $f(x, y, z) = x^2 + y^2 + 2z^2$ on the surface S defined by the equations $x + y + z = 3$ and $x - y + 3z = 2$. Using the first constraint find ϕ such that $z = \phi(x, y)$ on S. Now use the method of Lagrange multipliers to find the minimum of $f(x, y, \phi(x, y))$ on the level set $x - y + 3\phi(x, y) = 2$. Using both constraints express y and z as functions ϕ_1 and ϕ_2 of x for all points on S. Find the minimum of $f(x, \phi_1(x), \phi_2(x))$ using one-variable calculus.

3.6 Show that the maximum and minimum of $f(x, y, z) = x/a + y/b + z/c$ on the ellipsoid $(x/a)^2 + (y/b)^2 + (z/c)^2 = 1$ are $\sqrt{3}$ and $-\sqrt{3}$ respectively where a, b and c are positive constants.

3.7 Find the maximum value of xyz on the level set $F^{-1}(1)$ where

$$F(x, y, z) = \frac{1}{x} + \frac{1}{y} + \frac{1}{z} .$$

3.8 Use Lagrange multipliers to find the maximum volume of the rectangular solid in the first octant ($x \geq 0$, $y \geq 0$, $z \geq 0$) with one vertex at the origin and the opposite vertex lying in the plane $x/a + y/b + z/c = 1$ where a, b and c are positive constants.

3.9 Show that of all triangles inscribed in a fixed circle the equilateral triangle maximises

(a) the product of the lengths of the sides

(b) the sum of the squares of the lengths of the sides.

3.10 Find the minimum of $f(x, y, z) = 2y$ on the set $3x^2 - y^5 = 0$.

3.11 If $f : U(\text{open}) \subset \mathbb{R}^n \to \mathbb{R}$, $P \in U$ and $\nabla f(P) \neq 0$ find the direction, at P, in which f increases most rapidly.

3.12 Show that the maximum of $x_1 \cdots x_n$ on $\sum_{i=1}^{n} \frac{x_i^2}{i^2} = 1$ is $\frac{n!}{n^{n/2}}$.

3.13 Within a triangle there is a point P such that the sum of the squares of the distances to the sides is a minimum. Find this minimum in terms of the lengths of the sides and the area.

3.14 The straight line L in $\mathbb{R}^3$ is the intersection of the planes

$$\begin{aligned} x + y + 2z &= 5 \\ 2x - 3y + z &= 12 . \end{aligned}$$

Find the distance from the point $(10, 1, -6)$ to L.

3.15 What is the relationship between Exercise 2.13 and the Cauchy–Schwarz inequality (Example 3.4)?

4
Maxima and Minima on Open Sets

Summary. *We derive, using critical points and the Hessian, a method of locating local maxima, local minima and saddle points of a real-valued function defined on an open subset of $\mathbb{R}^n$.*

We turn to the problem of finding local maxima, local minima and saddle points of a real-valued function f on an open subset U of $\mathbb{R}^n$. A critical point is a *saddle point* if it is a local maximum when approached in one direction while it is a local minimum when approached in some other direction. The set of critical points of f on U, $\{X; \nabla f(X) = 0\}$, will include all local maxima, local minima and saddle points but may contain additional points.

The *Hessian* of f at $P \in U$, $H_{f(P)}$, is defined as the $n \times n$ matrix $\left(\dfrac{\partial^2 f}{\partial x_i \partial x_j}(P)\right)_{1 \leq i,j \leq n}$. To define the Hessian we are, of course, assuming that all first- and second-order partial derivatives of f exist. We use the convention that the order of differentiation is from right to left, i.e.

$$\frac{\partial^2 f}{\partial x_i \partial x_j} = \frac{\partial}{\partial x_i}\left(\frac{\partial f}{\partial x_j}\right).$$

If all these second-order partial derivatives exist and are continuous then the order of differentiation is immaterial and

$$\frac{\partial^2 f}{\partial x_i \partial x_j}(P) = \frac{\partial^2 f}{\partial x_j \partial x_i}(P)$$

for all i and j. In this case $H_{f(P)}$ is a symmetric $n \times n$ matrix. We will not prove this result but provide, in Exercise 4.6, a practical method which proves

35

that all functions you will probably ever encounter have this property. If $\mathbf{v} = (v_1, \ldots, v_n)$ is a vector in $\mathbb{R}^n$ and ${}^t\mathbf{v}$ is the corresponding column vector then

$$\frac{\partial f}{\partial \mathbf{v}}(P) = \sum_{i=1}^{n} v_i \frac{\partial f}{\partial x_i}(P)$$

and

$$
\begin{aligned}
\frac{\partial^2 f}{\partial \mathbf{v}^2}(P) = \frac{\partial}{\partial \mathbf{v}}\Big(\frac{\partial f}{\partial \mathbf{v}}\Big)(P) &= \sum_{i=1}^{n} v_i \frac{\partial}{\partial \mathbf{v}}\Big(\frac{\partial f}{\partial x_i}\Big)(P) \\
&= \sum_{i=1}^{n} v_i \Big(\sum_{j=1}^{n} v_j \frac{\partial^2 f}{\partial x_j \partial x_i}(P)\Big) \\
&= \sum_{i,j=1}^{n} v_i v_j \frac{\partial^2 f}{\partial x_i \partial x_j}(P) \\
&= \mathbf{v} H_{f(P)}{}^t\mathbf{v}.
\end{aligned}
$$

The main theoretical result on the existence of local maxima, local minima and saddle points is the following theorem.

Theorem 4.1

If U is an open subset of $\mathbb{R}^n$, $f: U \to \mathbb{R}$ is a twice continuously differentiable function on U and P is a critical point of f, i.e.

$$\nabla f(P) = \Big(\frac{\partial f}{\partial x_1}(P), \ldots, \frac{\partial f}{\partial x_n}(P)\Big) = 0$$

then

(1) f has a strict local maximum at P if and only if $\dfrac{\partial^2 f}{\partial \mathbf{v}^2}(P) < 0$ for all $\mathbf{v} \neq 0$,

(2) f has a strict local minimum at P if and only if $\dfrac{\partial^2 f}{\partial \mathbf{v}^2} > 0$ for all $\mathbf{v} \neq 0$,

(3) f has a saddle point at P if there exist $\mathbf{v}$ and $\mathbf{w}$ such that

$$\frac{\partial^2 f}{\partial \mathbf{v}^2}(P) < 0 < \frac{\partial^2 f}{\partial \mathbf{w}^2}(P).$$

To derive a practical test from this result we use linear algebra and Lagrange multipliers. To simplify matters we change our notation and let $\mathbf{v} = X = (x_1, \ldots, x_n)$, $a_{ij} = \dfrac{\partial^2 f}{\partial x_i \partial x_j}(P)$ and $A = (a_{ij})_{1 \leq i,j \leq n}$. With this notation

$$\frac{\partial^2 f}{\partial \mathbf{v}^2}(P) = \mathbf{v} H_{f(P)}{}^t\mathbf{v} = X A^t X.$$

Since

$$\frac{X}{\|X\|} A^t \left(\frac{X}{\|X\|}\right) = \frac{1}{\|X\|^2} X A^t X$$

for $X \neq 0$ we have

$$XA^tX > 0 \quad \text{for all } X \neq 0 \quad \Longleftrightarrow \quad \min_{\|X\|=1} XA^tX > 0$$

$$XA^tX < 0 \quad \text{for all } X \neq 0 \quad \Longleftrightarrow \quad \max_{\|X\|=1} XA^tX < 0$$

there exists $X, Y \in \mathbb{R}^n$ such that

$$XA^tX < 0 < YA^tY \quad \Longleftrightarrow \quad \min_{\|X\|=1} XA^tX < 0 < \max_{\|X\|=1} XA^tX.$$

We thus need to examine the extreme values of XA^tX on the set

$$\|X\|^2 = \langle X, X \rangle = \sum_{i=1}^{n} x_i^2 = 1.$$

Let

$$h(X) = h(x_1, \ldots, x_n) = \sum_{i,j=1}^{n} a_{ij} x_i x_j = XA^tX$$

and

$$g(X) = g(x_1, \ldots, x_n) = \sum_{i,j=1}^{n} x_i^2 = \langle X, X \rangle.$$

Since the set $g^{-1}(1)$ is compact and h is continuous the fundamental existence theorem implies that h has a maximum and minimum on $g^{-1}(1)$. Using the coordinate expansion of g we see that $\nabla g(X) = (2x_1, \ldots, 2x_n) = 2X$ and $\nabla g(X) \neq 0$ on g^{-1}. Hence we may apply the method of Lagrange multipliers to find the maximum and the minimum of h on the set $g^{-1}(1)$. For future reference we note that the level set $g^{-1}(1)$ is the unit sphere centred at the origin and the normal line to the sphere at the point X points in the *same* direction as X (Figure 4.1).

Similarly

$$\nabla h(X) = \left(\sum_{j=1}^{n} 2a_{j1} x_j, \ldots, \sum_{j=1}^{n} 2a_{jn} x_j\right) = 2XA.$$

By the method of Lagrange multipliers there exists at the maximum and minimum points of h on $g^{-1}(1)$ a real number λ such that

$$2XA = 2\lambda X.$$

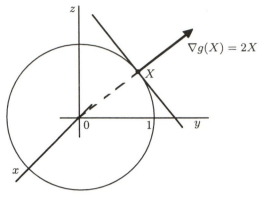

Figure 4.1.

Taking the transpose we get

$$A^tX = \lambda^t X = \lambda I^t X$$

i.e.

$$(A - \lambda I)^t X = 0. \tag{4.1}$$

Since $\|X\| = 1$, any λ which satisfies (4.1) is an *eigenvalue* of A and, moreover,

$$h(X) = X A^t X = \lambda X^t X = \lambda\langle X, X\rangle = \lambda.$$

Thus the maximum and minimum values of h are eigenvalues of A and are achieved at the corresponding unit eigenvectors. If all eigenvalues are positive then h is always positive and f has a local minimum at P, if all eigenvalues are negative then h is always negative and f has local maximum at P and if some are positive and some negative then h takes positive and negative values and f has a saddle point at P. This observation will play an important role in our study of Gaussian curvature in Chapter 16.

If λ is an eigenvalue of A the set

$$E_\lambda := \{X \in \mathbb{R}^n : A^tX = \lambda^tX\}$$

is a subspace of $\mathbb{R}^n$, called the λ-*eigenspace* of A, and the dimension of E_λ is called the *multiplicity* of the eigenvalue λ. An $n \times n$ symmetric matrix has n eigenvalues when eigenvalues are counted according to multiplicity, i.e. if E_λ is j–dimensional then λ is counted j times. We also have $\det(A) = \lambda_1 \cdots \lambda_n$.

For our test we just need to know the sign of the largest and smallest eigenvalues. Since eigenvalues may be difficult to calculate we will use a reasonably well-known result from linear algebra. If A is a square matrix then the $k \times k$

matrix A_k obtained by deleting all except the first k rows and k columns of A is called the $k \times k$ *principal minor* of A. We have $A_n = A$.

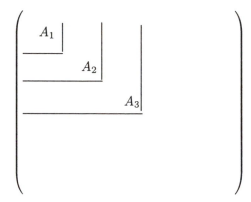

We require the following result:

if A is a symmetric $n \times n$ matrix then all eigenvalues of A are positive if and only if $\det(A_k) > 0$ for all k.

Proposition 4.1

If $f: U \subset \mathbb{R}^n \to \mathbb{R}$, U open, has continuous first- and second-order partial derivatives, P is a critical point of f, and $A = H_{f(P)}$ is the Hessian of f at P then the following hold:

(a) if $\det(A_{2k}) < 0$ for some k then P is a saddle point of f

(b) if $\det(A_n) \neq 0$ then

   (b1) f has a strict local minimum at P if and only if $\det(A_k) > 0$ for all k,

   (b2) f has a strict local maximum at P if and only if $(-1)^k \det(A_k) > 0$ for all k,

   (b3) f has a saddle point at P if and only if it has neither a strict local maximum nor a strict local minimum at P

(c) if $\det(A_n) = 0$ we call P a degenerate critical point of f (all other critical points are called non-degenerate) and higher order derivatives are often required to test the nature of the critical point at P.

Proof

(a) Suppose $\det(A_{2k}) < 0$ for some positive integer k, $2k \leq n$. If $P = (p_1, \ldots, p_n)$ let $Q = (p_1, \ldots, p_{2k})$. Consider the function $g: V \subset \mathbb{R}^{2k} \to \mathbb{R}$

defined by
$$g(x_1, \ldots, x_{2k}) = f(x_1, \ldots, x_{2k}, p_{2k+1}, \ldots, p_n)$$
where V is the open set in $\mathbb{R}^{2k}$ consisting of all $(x_1, \ldots, x_{2k})$ such that
$$(x_1, \ldots, x_{2k}, p_{2k+1}, \ldots, p_n) \in U.$$
It is easily checked that
$$H_{g(Q)} = \left(\frac{\partial^2 g}{\partial x_i \partial x_j}(Q) \right)_{1 \leq i,j \leq 2k} = \left(\frac{\partial^2 f}{\partial x_i \partial x_j}(P) \right)_{1 \leq i,j \leq 2k} = A_{2k}.$$
Let $\beta_1, \ldots, \beta_{2k}$ denote the $2k$ eigenvalues of the symmetric $2k \times 2k$ matrix A_{2k} counted according to multiplicity. We have
$$\beta_1 \cdots \beta_{2k} = \det(A_{2k}) < 0.$$
Since $2k$ is an even integer it follows that A_{2k} has positive and negative eigenvalues. Hence there exist $\mathbf{u} = (u_1, \ldots, u_{2k})$ and $\mathbf{v} = (v_1, \ldots, v_{2k})$ such that
$$\mathbf{u} A_{2k}{}^t\mathbf{u} < 0 < \mathbf{v} A_{2k}{}^t\mathbf{v}.$$
Let $\mathbf{w_1} = (u_1, \ldots, u_{2k}, 0, \ldots, 0)$ and $\mathbf{w_2} = (v_1, \ldots, v_{2k}, 0, \ldots, 0)$. Then
$$\begin{aligned}
\frac{\partial^2 f}{\partial \mathbf{w_1}^2}(P) &= \mathbf{w_1} H_{f(P)}{}^t\mathbf{w_1} \\
&= \mathbf{u} A_{2k}{}^t\mathbf{u} < 0 < \mathbf{v} A_{2k}{}^t\mathbf{v} \\
&= \mathbf{w_2} H_{f(P)}{}^t\mathbf{w_2} = \frac{\partial^2 f}{\partial \mathbf{w_2}^2}(P)
\end{aligned}$$
and f has a saddle point at P. This proves (a).

(b) The first part, (b1), follows directly from the linear algebra result quoted above. Since f has a local maximum at P if and only if $(-f)$ has a local minimum at the same point and $H_{-f(P)} = -H_{f(P)}$ it follows, from (b1), that f has a local maximum at P if $\det(-A_k) > 0$ for all k. Hence (b2) follows since $\det(-A_k) = (-1)^k \det(A_k)$. If (b1) and (b2) do not apply then, since all eigenvalues of A are non-zero, A has positive and negative eigenvalues and hence f has a saddle point at P. This completes the proof.

Proposition 4.1 enables us to classify all non-degenerate and some degenerate critical points. In practice the determinants, $\det(A_i)$, are calculated in the order $i = 1, 2, 3, \ldots$ and testing for *saddle points* is carried out as the calculations proceed. The critical point is a saddle point and the calculations stop when for the first time either of the following is observed:
$$\det(A_{2k}) < 0$$

$$\det(A_{2k-1}) \det(A_{2k+1}) < 0.$$

If the critical point is a local maximum or minimum then all the determinants must be calculated and then (b1), (b2) or (b3) will apply if $\det(A_n) \neq 0$.

For example suppose the Hessian at a critical point P is

$$A = \begin{pmatrix} 2 & 2 & 0 & 0 \\ 2 & 3 & 1 & 0 \\ 0 & 1 & -4 & 2 \\ 0 & 0 & 2 & -2 \end{pmatrix}$$

then $\det(A_1) = 2$, $\det(A_2) = 2$, $\det(A_3) = -10$ and $\det(A_1)\det(A_3) = -20 < 0$ implies that P is a saddle point.

Example 4.1

Let $f(x, y, z) = x^2y^2 - z^2 + 2x - 4y + z$. We have

$$\nabla f(x, y, z) = (2xy^2 + 2, 2x^2y - 4, -2z + 1).$$

If P is a critical point of f then

$$\begin{aligned} 2xy^2 + 2 &= 0 \\ 2x^2y - 4 &= 0 \\ -2z + 1 &= 0 \end{aligned}$$

Hence $z = 1/2$ from the third equation. From the first two equations we see that x and y are non-zero. Hence $xy^2 = -1$ and $x^2y = 2$ imply $xy^2/x^2y = -1/2 = y/x$ and $x = -2y$. We have $-2y \cdot y^2 = -1$, i.e. $y^3 = 1/2$ and $y = 2^{-1/3}$. From $x = -2y$ we obtain $x = -2^{2/3}$ and conclude that $(-2^{2/3}, 2^{-1/3}, 1/2)$ is the only critical point of f. A simple calculation shows

$$H_{f(x,y,z)} = \begin{pmatrix} 2y^2 & 4xy & 0 \\ 4xy & 2x^2 & 0 \\ 0 & 0 & -2 \end{pmatrix}$$

and

$$H_{f(-2^{2/3}, 2^{-1/3}, 1/2)} = \begin{pmatrix} 2^{1/3} & -4 \cdot 2^{1/3} & 0 \\ -4 \cdot 2^{1/3} & 2 \cdot 2^{4/3} & 0 \\ 0 & 0 & -2 \end{pmatrix}.$$

Since $\det(2^{1/3}) > 0$ and

$$\begin{aligned} \det \begin{pmatrix} 2^{1/3} & -4 \cdot 2^{1/3} \\ -4 \cdot 2^{1/3} & 2 \cdot 2^{4/3} \end{pmatrix} &= 2 \cdot 2^{5/3} - 16 \cdot 2^{2/3} \\ &= 4 \cdot 2^{2/3} - 16 \cdot 2^{2/3} < 0 \end{aligned}$$

the critical point $(-2^{2/3}, 2^{-1/3}, 1/2)$ is a saddle point of f.

Example 4.2

We wish to find and classify the non-degenerate critical points of $f(x,y,z) = x^2y + y^2z + z^2 - 2x$. We have

$$\nabla f(x,y,z) = (2xy - 2, x^2 + 2yz, y^2 + 2z)$$

and the critical points satisfy the equations

$$2xy - 2 = 0, \quad x^2 + 2yz = 0 \quad \text{and} \quad y^2 + 2z = 0.$$

Substituting $z = -y^2/2$ into the second equation implies $y^3 = x^2$. Hence, the first equation shows $y^{5/2} = 1$ and we have $y = 1$ and $z = -1/2$. From $xy = -1$ we get $x = 1$ and $(1, 1, -1/2)$ is the only critical point of f. We have

$$H_{f(x,y,z)} = \begin{pmatrix} 2y & 2x & 0 \\ 2x & 2z & 2y \\ 0 & 2y & 2 \end{pmatrix}$$

and

$$H_{f(1,1,-1/2)} = \begin{pmatrix} 2 & 2 & 0 \\ 2 & -1 & 2 \\ 0 & 2 & 2 \end{pmatrix}.$$

Since $\det(2) > 0$ and

$$\det \begin{pmatrix} 2 & 2 \\ 2 & -1 \end{pmatrix} = -2 - 4 < 0$$

the point $(1, 1, -1/2)$ is a saddle point of f.

EXERCISES

4.1 Classify the non-degenerate critical points of

(a) $x^2 + xy + 2x + 2y + 1$
(b) $x^3 + y^3 - 3xy$
(c) $(x-1)e^{xy}$
(d) $(2-x)(4-y)(x+y-3)$
(e) $4xyz - x^4 - y^4 - z^4$
(f) $xyze^{-x^2-y^2-z^2}$
(g) $xy + x^2z - x^2 - y - z^2$.

4.2 If $a > b > c > 0$ show that the function

$$f(x,y,z) = (ax^2 + by^2 + cz^2)e^{-x^2-y^2-z^2}$$

has two local maxima, one local minimum and four saddle points. By using the fundamental existence theorem, as in Example 3.3, show that f has a maximum and minimum on $\mathbb{R}^3$.

4.3 Show that the function $xyz(x + y + z - 1)$ has one non-degenerate critical point and an infinite set of degenerate critical points. Show that the non-degenerate critical point is a local minimum.

4.4 Show that every critical point of $\dfrac{x^3 + y^3 + z^3}{xyz}$ is degenerate.

4.5 Find the distance from the point $(-1, 1, 1)$ to the level set $z = xy$.

4.6 Let $U = (a_1, b_1) \times (a_2, b_2) \times \cdots \times (a_n, b_n) \subset \mathbb{R}^n$ and let

$$M(U) \;=\; \{f\colon U \to \mathbb{R} : \text{ all first- and second-order partial}$$
$$\text{derivatives of } f \text{ exist and } \frac{\partial^2 f}{\partial x_i \partial x_j}(P) = \frac{\partial^2 f}{\partial x_j \partial x_i}(P)$$
$$\text{for all } P \in U \text{ and all } i,\, j,\, 1 \le i, j \le n\}.$$

Show that $M(U)$ has the following properties:

(i) if $f,\, g \in M(U)$ and $c \in \mathbb{R}$ then $f \pm g$, $f \cdot g$ and $c \cdot f \in M(U)$ and if $g \ne 0$ then $f/g \in M(U)$

(ii) e_i^*, $i = 1, \ldots, n$, belong to $M(U)$

(iii) if $f \in M(U)$ and $\phi\colon \mathbb{R} \to \mathbb{R}$ is twice continuously differentiable then $\phi \circ f \in M(U)$.

Using (i), (ii), (iii) and Exercise 1.20 show that

$$h(x, y, z) = \sin^2\left(\frac{e^{xyz}}{y^2 + z^2 + 1}\right)$$

lies in $M(U)$. Verify this result by calculating directly the appropriate second-order partial derivatives of h.

4.7 Let $X_1, \ldots, X_m$ be m points in $\mathbb{R}^n$. Show that $\sum_{i=1}^m \|X - X_i\|^2$ achieves its absolute minimum at $X = \frac{1}{m} \sum_{i=1}^m X_i$. Interpret your result geometrically.

4.8 If $z = \phi(x, y)$ satisfies the equation

$$x^2 + 2y^2 + 3z^2 - 2xy - 2yz = 2$$

find the points (x, y) at which ϕ has a local maximum or a local minimum.

4.9 If A is a symmetric matrix and X and Y are eigenvectors corresponding to different eigenvalues show that $\langle X, Y \rangle = 0$.

5
Curves in $\mathbb{R}^n$

Summary. *We introduce and discuss the concept of directed curve in $\mathbb{R}^n$. We obtain a formula for the length of a curve, prove the existence of unit speed parametrizations and define piecewise smooth curves.*

Directed and parametrised curves play a role in many of the topics discussed in the remaining chapters of this book, e.g. line integrals, existence of a potential, Stokes' theorem and the geometry of surfaces in $\mathbb{R}^3$, and, furthermore, in a simple fashion introduce us to concepts such as parametrizations and orientations that are later developed and generalised in more involved settings.

We begin by giving a rigorous definition of directed curve. This may appear complicated and unnecessarily cumbersome at first glance and so we feel it proper to elaborate on why each condition is included. It is always important in mathematics to understand the basic *definition* and to refer to it until one appreciates each part separately and the totality of parts collectively. As progress is achieved there is usually less need to refer to the definition but in case of ambiguity the definition is the book of *rules*. The only requirement in a definition is that it be consistent, i.e. that the various conditions do not contradict one another or the rules of mathematics. Apart from this there is freedom in the choice of conditions and, indeed, many books would give a slightly different definition of directed curve. The differences depend on the degree of generality sought, the results aimed at and the methods used. However, all definitions of curve, directed curve and parametrised curve contain the same essential features.

Definition 5.1

A *directed* (or *oriented*) curve in $\mathbb{R}^n$ is a quadruple $\{\Gamma, A, B, \mathbf{v}\}$ where Γ is a set of points in $\mathbb{R}^n$; A and B are points in Γ, called respectively the *initial* and *final points* of Γ; $\mathbf{v}$ is a unit vector in $\mathbb{R}^n$ called the *initial direction*, for which there exists a mapping $P: [a, b] \to \mathbb{R}^n$, called a *parametrization* of Γ, such that the following conditions hold:

(a) there exists an *open* interval I, containing $[a, b]$ and a mapping from I into $\mathbb{R}^n$ which has derivatives of all orders and which coincides with P on $[a, b]$

(b) $P([a, b]) = \Gamma$, $P(a) = A$, $P(b) = B$ and $P'(a) = \alpha \mathbf{v}$ for some $\alpha > 0$

(c) $P'(t) \neq 0$ for all $t \in [a, b]$

(d) P is injective (i.e. one to one) on $[a, b)$ and $(a, b]$.

Condition (a) is a rather strong regularity condition – we require the first derivative to define the length, the second to define curvature, the third to define torsion and the fourth to obtain the Frenet–Serret equations, and at this stage we felt it was just as easy to assume that we had derivatives of all orders. We also wished to have derivatives at the end points of the interval $[a, b]$ and for this reason we assumed that P has an extension as a *smooth* function (i.e. as a function with derivatives of all orders) to an open interval containing $[a, b]$. We could achieve precisely the same degree of smoothness by using one-sided derivatives at a and b but felt this *appears* even more complicated. All definitions of a curve will include, as an essential feature a *continuous* mapping P from an interval I in $\mathbb{R}$ *onto* Γ. The degree of differentiability, whether the interval I is open, closed, finite or infinite may be regarded as options that are available. We have chosen the options that suit our purposes.

The essential feature of condition (b), as we have just noted, is $P([a, b]) = \Gamma$. The remaining parts endow the set Γ with a sense of direction. If $A \neq B$ then the conditions $P(a) = A$ and $P(b) = B$ define a direction along Γ and in this case the condition $P'(a) = \alpha \mathbf{v}$ is redundant as $\mathbf{v}$ is determined by $\{\Gamma, A, B\}$ (Figure 5.1).

Figure 5.1.

If, however, $A = B$ then we have a *closed curve*, and it is necessary to distinguish between the two directions we may travel around Γ. In the case of curves in $\mathbb{R}^2$ we have clockwise and anticlockwise or counterclockwise directions. In $\mathbb{R}^n$ we do not have such a concept and instead specify the direction along the curve by giving an initial direction $\mathbf{v}$ (Figure 5.2).

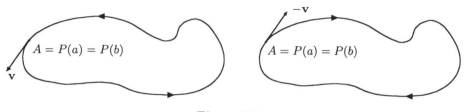

Figure 5.2.

Note that when we know Γ and A then $\mathbf{v} = \pm P'(a) \,/\, \|P'(a)\|$ and the condition $\alpha > 0$ distinguishes between the two signs and fixes the direction. Condition (c) is necessary to obtain a unit speed parametrization and we have already used $P'(a) \neq 0$ in (b). This condition excludes curves with corners but we get around this problem by defining piecewise smooth curves. These are obtained by placing end to end a finite number of directed curves. The transition from directed curves to piecewise smooth curves is painless. Later we will require $P''(t) \neq 0$ in order to define a unit normal to a directed curve in $\mathbb{R}^3$.

Since we allow $A = B$ it follows that the mapping P may not be injective on $[a, b]$. However, we do not wish the curve to cross itself (Figure 5.3(a)) or to half cross itself (Figure 5.3(b)) as these lead to unnecessary complications and we have included condition (d) to exclude such possibilities.

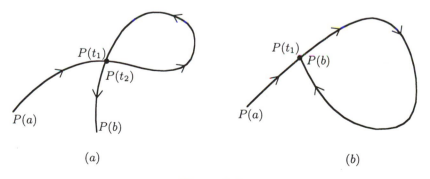

(a)   (b)

Figure 5.3.

A continuous mapping $P\colon [a,b] \to \mathbb{R}^n$ which satisfies (a), (c) and (d) is called a *parametrised curve*. A parametrised curve determines precisely one directed curve

$$\left\{ P([a,b]),\ P(a),\ P(b),\ \frac{P'(a)}{\|P'(a)\|} \right\}$$

for which it is a parametrization.

Since the terminology "directed curve" and the notation $\{\Gamma, A, B, \mathbf{v}\}$ are rather cumbersome we will use the term *curve* and the notation Γ in all cases where there is little danger of confusion. If $A \neq B$ we sometimes write $\{\Gamma, A, B\}$. If we need to use coordinates we usually let $P(t) = \big(x_1(t), \ldots x_n(t)\big)$ if $\Gamma \in \mathbb{R}^n$ and, when $n = 3$, we let $P(t) = \big(x(t), y(t), z(t)\big)$. It is helpful to think of $[a,b]$ as an interval of time and $P(t)$ as the position of a particle at time t as it travels along the route Γ from A to B.

We call $P'(t)$ the *velocity* and $\|P'(t)\|$ the *speed* at time t. Since *distance=* *speed×time* the formula

$$l(\Gamma) = \int_a^b \|P'(t)\|\, dt$$

where $l(\Gamma)$ is the length of Γ, is not surprising. We shall, however, pause to prove this formula in order to show the usefulness of vector notation. Since P is differentiable we have for all t and $t + \Delta t$ in $[a,b]$

$$P(t + \Delta t) = P(t) + P'(t)\Delta t + g(t, \Delta t) \cdot \Delta t$$

where $g(t, \Delta t) \to 0$ as $\Delta t \to 0$ for any fixed t. Hence $P(t+\Delta t) - P(t) \approx P'(t)\Delta t$ for Δt close to zero. If we partition $[a,b]$ we get a corresponding partition of Γ and an approximation of the length of Γ (Figure 5.4).

We have

$$
\begin{aligned}
l(\Gamma) \quad &\approx \quad \sum_i \|P(t_{i+1}) - P(t_i)\| \\
&\approx \quad \sum_i \|P'(t_i)\|\Delta t_i \\
&\longrightarrow \quad \int_a^b \|P'(t)\|\, dt
\end{aligned}
$$

as we take finer and finer partitions of $[a,b]$. In terms of coordinates we have $P(t) = \big(x_1(t), \ldots, x_n(t)\big)$, $P'(t) = \big(x_1'(t), \ldots, x_n'(t)\big)$ and

$$l(\Gamma) = \int_a^b \|P'(t)\|\, dt = \int_a^b \big(x_1'(t)^2 + \cdots + x_n'(t)^2\big)^{1/2} dt.$$

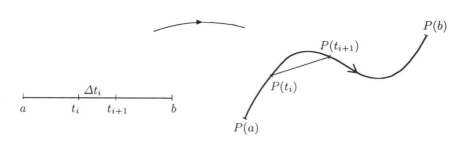

Figure 5.4.

Example 5.1

Let $P(t) = \cos t\mathbf{i} + \sin t\mathbf{j} + t\mathbf{k}$, $t \in [0, 2\pi]$, denote a parametrised curve Γ in $\mathbb{R}^3$, where $\mathbf{i}$, $\mathbf{j}$ and $\mathbf{k}$ denote the standard unit vector basis in $\mathbb{R}^3$, i.e. $\mathbf{i} = (1,0,0)$, $\mathbf{j} = (0,1,0)$ and $\mathbf{k} = (0,0,1)$. The curve Γ is part of a *helix* – it *spirals* around a vertical *cylinder* (Figure 5.5).

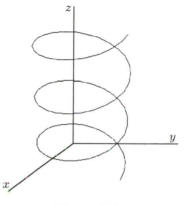

Figure 5.5.

If we consider only the first two coordinates, this amounts to projecting onto the $\mathbb{R}^2$ plane in $\mathbb{R}^3$, we get the standard parametrization $t \to (\cos t, \sin t)$ of the unit circle. Hence, disregarding the final coordinate, which we take to be the height, the particle appears to move in a circle. We have

$$P'(t) = -\sin t\mathbf{i} + \cos t\mathbf{j} + \mathbf{k}.$$

The rate of change of the height is given by the coefficient of the $\mathbf{k}$ term and so the particle is rising with constant speed. We have

$$\|P'(t)\| = \left((-\sin t)^2 + (\cos t)^2 + 1^2\right)^{1/2} = \sqrt{2}$$

and

$$l(\Gamma) = \int_0^{2\pi} \sqrt{2}dt = 2\sqrt{2}\pi.$$

If $P\colon [a, b] \to \mathbb{R}^n$ is a parametrization of the directed curve Γ we define the *length function* by the formula

$$s(t) = \int_a^t \|P'(x)\|dx$$

for all $t \in [a, b]$. If $l = l(\Gamma)$ then $s\colon [a, b] \to [0, l]$ and, by the one-variable fundamental theorem of calculus, $s'(t) = \|P'(t)\| > 0$. Hence s is strictly increasing, $s^{-1}\colon [0, l] \to [a, b]$ has derivatives of all orders on $[0, l]$ and $P \circ s^{-1}$ maps $[0, l]$ onto Γ (Figure 5.6). For the inverse function s^{-1} we have

$$(s^{-1})'(t) = \frac{1}{s'(s^{-1}(t))} = \frac{1}{\|P'(s^{-1}(t))\|}$$

and hence

$$\|(P \circ s^{-1})'(t)\| = \frac{\|P'(s^{-1}(t))\|}{s'(s^{-1}(t))} = \frac{\|P'(s^{-1}(t))\|}{\|P'(s^{-1}(t))\|} = 1.$$

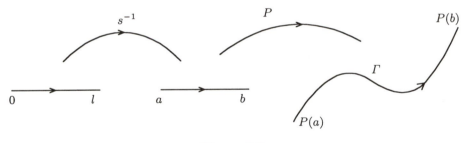

Figure 5.6.

Since the remaining conditions (for a parametrization) are easily checked it follows that $P \circ s^{-1}$ is a parametrization of Γ and we have proved the following result.

Proposition 5.1

Directed curves admit unit speed parametrizations.

If two particles start at the point A at time zero and both proceed along Γ towards B at unit speed then their positions on Γ at time t will always agree.

This shows that a directed curve of length l admits a *unique* unit speed parametrization on $[0, l]$. Using our construction of unit speed curve we may restate this as follows: if $P_1 : [a, b] \to \Gamma$ and $P_2 : [c, d] \to \Gamma$ are any two parametrizations of the directed curve Γ, of length l, and

$$s_1(t) = \int_a^t \|P_1'(x)\| dx, \qquad s_2(t) = \int_c^t \|P_2'(x)\| dx$$

are the associated length functions then

$$P_1 \circ s_1^{-1}(t) = P_2 \circ s_2^{-1}(t)$$

for all $t \in [0, l]$.

To include curves with corners we extend the concept of directed curve to that of piecewise smooth directed curve. A finite collection of directed curves $\{\Gamma_i, A_i, B_i, \mathbf{v_i}\}_{i=1}^n$ is called a *piecewise smooth directed curve* if

(a) each $\{\Gamma_i, A_i, B_i, \mathbf{v_i}\}$ is a directed curve,

(b) $B_i = A_{i+1}$ for $i = 1, \ldots, n-1$ (i.e. the *final* point of Γ_i coincides with the *initial* point of Γ_{i+1}).

If $B_n = A_1$ we say that the piecewise smooth directed curve is *closed*. We use the notation Γ for a piecewise smooth directed curve and A and B for its initial and final points respectively. This definition is rather general and apart from curves with corners it also includes curves which cross one another. In general such curves do not admit a unit speed parametrization but it can be shown that there exists a *continuous* parametrization

$$P : [a, b] \longrightarrow \Gamma = \bigcup_{i=1}^n \Gamma_i$$

and a partition $\{a_0 = a, a_1, a_2, \ldots, a_n = b\}$ of $[a, b]$ such that $P\big([a_{i-1}, a_i]\big) = \Gamma_i$, $P(a_{i-1}) = A_i$, $P(a_i) = B_i$, $i = 1, \ldots, n$, P has derivatives of all orders on $[a, b]$ and $P'(t) \neq 0$ for $t \neq a_0, \ldots, a_n$.

A piecewise smooth directed curve $\Gamma = \{\Gamma_i, A_i, B_i, \mathbf{v_i}\}_{i=1}^n$ is studied and applied by considering each of the component sections $\{\Gamma_i, A_i, B_i, \mathbf{v_i}\}$ in turn (see for instance Exercise 5.6 and our method of finding a potential in the next chapter).

EXERCISES

5.1 Find the length of the curve parametrised by
$$P(t) = (2\cosh 3t, -2\sinh 3t, 6t), \quad 0 \le t \le 5.$$

5.2 Show that the following parametrizations are unit speed

(a) $P_1(s) = \frac{1}{2}\left(s + \sqrt{s^2+1}, (s+\sqrt{s^2+1})^{-1}, \sqrt{2}\log(s+\sqrt{s^2+1})\right)$,
$s \in [0,1]$

(b) $P_2(s) = \left(\dfrac{(1+s)^{3/2}}{3}, \dfrac{(1-s)^{3/2}}{3}, \dfrac{s}{\sqrt{2}}\right), \quad s \in [-1, +1]$

(c) $P_3(s) = \dfrac{1}{2}\left(\cos^{-1}(s) - s\sqrt{1-s^2}, 1 - s^2, 0\right), \quad s \in [0,1].$

5.3 Let r and h denote positive numbers. Find a unit speed parametrization of the helix
$$P(t) = (r\cos t, r\sin t, ht), \quad 0 \le t \le 6\pi.$$

5.4 Obtain unit speed parametrizations of the curves defined by

(a) $t \longrightarrow (e^t \cos t, e^t \sin t, e^t), \quad t \in [a, b]$

(b) $t \longrightarrow (\cosh t, \sinh t, t), \quad t \in [a, b].$

5.5 Parametrise the curve of intersection of the sphere $x^2 + y^2 + z^2 = 16$ and the cylinder $x^2 + (y-2)^2 = 4$ which lies in the first octant.

5.6 Parametrise the anticlockwise directed triangle in $\mathbb{R}^2$ with vertices $(1, 2)$, $(-1, -2)$ and $(4, 0)$ as a piecewise smooth curve.

5.7 Find the closest points on the curve $x^2 - y^2 = 1$ to $(a, 0)$ where (i) $a = 4$ (ii) $a = 2$ (iii) $a = \sqrt{2}$.

5.8 Let f denote a real-valued differentiable function defined on an open subset U of $\mathbb{R}^n$ and suppose $\nabla f(X) \neq 0$ for all X in U. Let P denote a parametrised curve in U. Use the chain rule to show that $\dfrac{d}{dt}(f \circ P)(t) = \langle \nabla f(P(t)), P'(t) \rangle$ and hence deduce, using the Cauchy–Schwarz inequality, that $\nabla f(X_0)$ gives the direction of maximum increase of f at X_0. Show that $\|\nabla f(X_0)\|$ is the maximum rate of increase.

If Γ is a level set of f, parameterised by P, show that $\nabla f(P(t)) \perp P'(t)$ for all t. Hence find, when $n = 2$, the tangent line and the normal line to the level set of f, which passes through X_0, at X_0.

6
Line Integrals

Summary. *We integrate vector-valued and scalar-valued functions along a directed curve in $\mathbb{R}^n$. We discuss scalar and vector potentials and define the curl of a vector field in $\mathbb{R}^3$.*

The differential calculus was developed to study extremal (i.e. maximal and minimal) values of functions. Since it is only possible to discuss the maximum and minimum of a real-valued function it is not surprising that such functions occupy a prominent role in several-variable differential calculus. However, in moving to integration theory it is more natural (and more natural in mathematics usually means more useful, more efficient and more elegant) to consider vector-valued functions where the domain and the range share, in perhaps a loose way, a *common dimension*. Formally, we have the following definition of a *vector field*.

Definition 6.1

A function F which maps a subset U of $\mathbb{R}^n$ into $\mathbb{R}^n$ is called a vector field on U.

If U is an open subset of $\mathbb{R}^n$ and the vector field has derivatives of all orders we call it a *smooth vector field* and if U is arbitrary and the vector field is continuous we use the term *continuous vector field*. We shall also use the notation **F** to denote a vector field.

The gradient is an important example of a vector field, i.e. if U is open and

$f : U \subset \mathbb{R}^n \to \mathbb{R}$ is differentiable then $\nabla f : U \to \mathbb{R}^n$ is a vector field on U.

Another useful example occurs when Γ is a directed curve in $\mathbb{R}^n$ and F is a function which assigns a vector in $\mathbb{R}^n$ to each point on Γ – in this case we say that F is a vector field *along* Γ. For example, if P is a parametrization of a directed curve Γ in $\mathbb{R}^n$ then the mapping

$$P(t) \in \Gamma \to P'(t) \in \mathbb{R}^3$$

is a smooth vector field along Γ.

Vector fields, which assign vectors to points in the domain of definition, are often represented as in Figure 6.1.

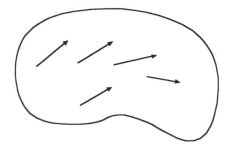

Figure 6.1.

This representation is useful in locating zeros and suggesting properties such as continuity and smoothness. Furthermore, it allows various physical interpretations of vector fields, e.g. as the velocity of a moving fluid and the flow of an electric current which lead, in turn, to important physical and engineering applications.

We begin our study of integration theory by defining the integral of a vector field F along a directed curve Γ. Let $P : [a, b] \to \Gamma$ denote a parametrization of Γ. To each partition of $[a, b]$ we obtain a partition of Γ (Figure 6.2) and the Riemann sum

$$\sum_i F\big(P(t_i)\big) \cdot \big(P(t_{i+1}) - P(t_i)\big) \approx \sum_i F\big(P(t_i)\big) \cdot P'(t_i)\Delta t_i$$

where $\Delta t_i = t_{i+1} - t_i$ and $\cdot$ denotes the inner product in $\mathbb{R}^n$. Note that we are using, as usual, the linear approximation to $P(t + \Delta t)$, $P(t) + P'(t)\Delta t$. If F is a continuous vector field along Γ, i.e. if the mapping $t \in [a, b] \to F(P(t))$ is continuous, then as we take finer and finer partitions of $[a, b]$ the Riemann sums converge to the limit

$$\int_a^b F\big(P(t)\big) \cdot P'(t)dt.$$

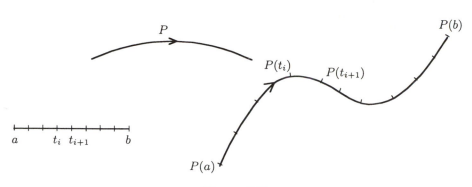

Figure 6.2.

We denote this integral by $\int_\Gamma F$, since we shall shortly prove that it is independent of the parametrization P, and call it the *line integral* of F over Γ. In terms of coordinates, for instance in the case $n = 3$, we have

$$F = (f, g, h), \quad P(t) = \big(x(t), y(t), z(t)\big), \quad P'(t) = \big(x'(t), y'(t), z'(t)\big)$$

and

$$\int_\Gamma F = \int_a^b \Big[f\big(x(t), y(t), z(t)\big) \cdot x'(t) + g\big(x(t), y(t), z(t)\big) \cdot y'(t)$$
$$+ h\big(x(t), y(t), z(t)\big) \cdot z'(t) \Big] dt .$$

This is frequently written in the form

$$\int_\Gamma f\, dx + g\, dy + h\, dz.$$

If $(\Gamma, A, B, \mathbf{v})$ is a directed curve and $\widetilde{\Gamma}$ is obtained by changing the direction along the curve Γ, i.e. $\widetilde{\Gamma} = (\Gamma, B, A, -\mathbf{v})$, we define $\int_{\widetilde{\Gamma}} F$ as $-\int_\Gamma F$ for any continuous vector field F.

Example 6.1

We evaluate

$$\int_\Gamma xy\, dx + xz^2\, dy + xyz\, dz$$

where the curve Γ is parametrised by

$$P(t) = (t, t^2, t^3), \quad 0 \le t \le 1.$$

In coordinates we have

$$
\begin{cases}
x = x(t) = t \\
y = y(t) = t^2 \\
z = z(t) = t^3
\end{cases}
\implies
\begin{cases}
\dfrac{dx}{dt} = x'(t) = 1 \\
\dfrac{dy}{dt} = y'(t) = 2t \\
\dfrac{dz}{dt} = z'(t) = 3t^2
\end{cases}
\implies
\begin{cases}
dx = dt \\
dy = 2t\,dt \\
dz = 3t^2 dt
\end{cases}
$$

and

$$
\begin{aligned}
\int_\Gamma xy\,dx + xz^2\,dy + xyz\,dz &= \int_0^1 t^3 dt + t^7 \cdot 2t\,dt + t^6 \cdot 3t^2 dt \\
&= \left[\frac{t^4}{4} + \frac{2t^9}{9} + \frac{3t^9}{9} \right]_0^1 = \frac{29}{36}.
\end{aligned}
$$

Alternatively, changing to vector notation, we let $F(x, y, z) = (xy, xz^2, xyz)$. Then $F(P(t)) = (t^3, t^7, t^6)$ and $P'(t) = (1, 2t, 3t^2)$. Hence

$$
\begin{aligned}
\int_\Gamma F &= \int_0^1 F(P(t)) \cdot P'(t)dt = \int_0^1 (t^3, t^7, t^6) \cdot (1, 2t, 3t^2)dt \\
&= \int_0^1 (t^3 + 2t^8 + 3t^8)dt = \frac{29}{36}.
\end{aligned}
$$

We return to the general situation. If $P\colon [a, b] \to \Gamma$ and $Q\colon [c, d] \to \Gamma$ are two parametrizations of Γ with length functions s and s_1 respectively, then, as we saw in Chapter 5,

$$
P \circ s^{-1} = Q \circ s_1^{-1} \quad \text{on } [0, l], \quad l = \text{length of } \Gamma.
$$

Using the one-variable change of variables, $y = s^{-1}(t)$ and $x = s_1^{-1}(t)$, we obtain

$$
\begin{aligned}
\int_0^l &F(P \circ s^{-1}(t)) \cdot (P \circ s^{-1})'(t)dt \\
&= \int_0^l F(P \circ s^{-1}(t)) \cdot P'(s^{-1}(t))(s^{-1})'(t)dt \\
&= \int_a^b F(P(y)) \cdot P'(y)dy.
\end{aligned}
$$

Similarly

$$
\int_0^l F(Q \circ s_1^{-1}(t)) \cdot (Q \circ s_1^{-1})'(t)dt = \int_c^d F(Q(x)) \cdot Q'(x)dx.
$$

Since $P \circ s^{-1} = Q \circ s_1^{-1}$ this implies

$$\int_a^b F\big(P(y)\big) \cdot P'(y) dy = \int_c^d F\big(Q(x)\big) \cdot Q'(x) dx$$

and we get the same value no matter which parametrization is used. This justifies the notation $\int_\Gamma F$.

If $P: [a,b] \to \Gamma$ is a parametrization of the directed curve Γ in $\mathbb{R}^n$ and $t \in [a,b]$ let $T(t) = P'(t) / \|P'(t)\|$. We call $T(t)$ the *unit tangent* to the (directed) curve at $P(t)$. It is easily seen that any two parametrizations define the same unit tangent vector at each point of Γ. This leads to another way of writing line integrals. If F is a vector field along Γ (always of course assumed to be continuous) then

$$\begin{aligned}
\int_\Gamma F &= \int_a^b F\big(P(t)\big) \cdot P'(t) dt \\
&= \int_a^b F\big(P(t)\big) \cdot \frac{P'(t)}{\|P'(t)\|} \|P'(t)\| dt \\
&= \int_a^b F\big(P(t)\big) \cdot T(t) \|P'(t)\| dt \\
&= \int_a^b (F \cdot T) ds
\end{aligned}$$

and when written in this form one should remember, in applying a parametrization P, that F and T are both evaluated at $P(t)$ and $ds = \|P'(t)\| dt$.

Real-valued functions or *scalar fields*, such as the speed of a parametrization, can also be defined along a directed curve Γ. Since these are not endowed with a sense of direction we cannot apply directly our definition of integral. Fortunately, we have just observed a *special* or *privileged* direction associated with each point on a curve, the tangent direction, and by associating the continuous real-valued function $f: \Gamma \to \mathbb{R}$ with the vector field $fT: \Gamma \to \mathbb{R}^n$ we can define $\int_\Gamma f$. The privileged direction on an oriented surface in $\mathbb{R}^3$ is the *normal* direction and in this case it is also possible to consider scalar-valued integration as a special case of vector-valued integration.

If $P: [a,b] \to \Gamma$ is a parametrization of the directed curve Γ we define

$$\begin{aligned}
\int_\Gamma f &= \int_\Gamma fT = \int_a^b f\big(P(t)\big) T(t) \cdot T(t) \|P'(t)\| dt \\
&= \int_a^b f\big(P(t)\big) \|P'(t)\| dt \qquad\qquad (6.1)
\end{aligned}$$

Because of (6.1) we sometimes write $\int_\Gamma f ds$ in place of $\int_\Gamma f$.

We now seek to identify those vector fields in $\mathbb{R}^n$ which are the gradient of a scalar-valued function. Vector fields of this kind are called *conservative* and are said to have a scalar potential. If $\nabla f = F$ we call f a *scalar potential* of F. Our investigation of this problem leads to a generalisation of the *fundamental theorem of one-variable calculus*

$$\int_a^b g'(t)dt = g(b) - g(a)$$

where g is continuous on $[a, b]$ and differentiable on (a, b).

We begin by considering properties of the gradient of $f: U \subset \mathbb{R}^n \to \mathbb{R}$. If Γ is a directed curve in U parametrised by $P: [a, b] \to \mathbb{R}^n$ then

$$\begin{aligned}
\int_\Gamma \nabla f &= \int_a^b \nabla f\big(P(t)\big) \cdot P'(t)dt \\
&= \int_a^b \frac{d}{dt}(f \circ P)(t)dt \quad \text{(chain rule)} \\
&= \Big[f \circ P(t) \Big]_a^b \quad \text{(fundamental theorem of calculus in } \mathbb{R}\text{)} \\
&= f\big(P(b)\big) - f\big(P(a)\big).
\end{aligned}$$

Thus the line integral of ∇f along Γ depends only on the values of f at the initial and final points of the curve. This, as we shall see in Proposition 6.1, characterises vector fields which have a scalar potential. If Γ is a piecewise smooth directed curve in $\mathbb{R}^n$ which is the union of directed curves $(\Gamma_i)_{i=1}^k$ and F is either a vector field along Γ or a real-valued function on Γ we let

$$\int_\Gamma F = \sum_{i=1}^k \int_{\Gamma_i} F.$$

In the next proposition it is necessary to assume that any pair of points in the open set U can be joined by a piecewise smooth directed curve which lies in U; an open set of this kind is said to be *connected*.

Proposition 6.1

Let $F: U \subset \mathbb{R}^n \to \mathbb{R}^n$ denote a continuous vector field on the connected open subset U of $\mathbb{R}^n$. If for any two points A and B in $\mathbb{R}^n$ and any two piecewise smooth directed curves Γ_1 and Γ_2 joining A and B we have

$$\int_{\Gamma_1} F = \int_{\Gamma_2} F$$

then F has a potential.

Proof

Let A denote a fixed point in U. For any X in U let $f(X) = \int_\Gamma F$ where Γ is any piecewise smooth directed curve in U joining A to X. By our hypothesis f is well defined, i.e. there is no ambiguity in the definition. Let $F = (f_1, \ldots, f_n)$ and let Γ denote a curve joining A to X. Fix j, $1 \leq j \leq n$, and let $Y = he_j$ where h denotes a real number close to 0. Let Γ_1 denote the directed curve parametrised by $P(t) = X + tY$, $0 \leq t \leq 1$. Then Γ_1 joins X to $X + Y$ and $\Gamma \cup \Gamma_1$ is a piecewise smooth directed curve in U which joins A to $X + Y$ (Figure 6.3).

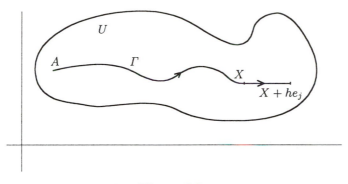

Figure 6.3.

Hence

$$f(X + Y) - f(X) = \int_{\Gamma \cup \Gamma_1} F - \int_\Gamma F$$

$$= \int_{\Gamma_1} F = \int_0^1 F\big(P(t)\big) \cdot P'(t)dt.$$

Since $P'(t) = Y = he_j$

$$f(X + Y) - f(X) = \int_0^1 F\big(P(t)\big) \cdot he_j dt$$

$$= h \int_0^1 f_j(X + tY)dt$$

$$= hf_j(X) + h \int_0^1 \Big(f_j(X + tY) - f_j(X)\Big)dt.$$

As F is continuous, each component of F and, in particular, f_j is also continuous. Hence

$$\max_{0 \leq t \leq 1} \big|f_j(X + tY) - f_j(X)\big| \longrightarrow 0 \quad \text{as } h \to 0$$

and

$$\left| \int_0^1 \Big(f_j(X + tY) - f_j(X) \Big) dt \right| \leq \max_{0 \leq t \leq 1} \big| f_j(X + tY) - f(X) \big|.$$

We have shown

$$f(X + he_j) = f(X) + hf_j(X) + g(X, h)h$$

where $g(X, h) \to 0$ as $h \to 0$. Hence $\dfrac{\partial f}{\partial x_j}(X) = f_j(X)$ and this completes the proof.

Proposition 6.1 can be used to *find* potentials but is not very practical for *showing* existence. We need a simpler method which follows from the observation:

if $F = (f_1, \ldots, f_n)$ is a continuously differentiable vector field with potential f, i.e. $f_j = \dfrac{\partial f}{\partial x_j}$ for all j, then for all i and j

$$\frac{\partial f_i}{\partial x_j} = \frac{\partial}{\partial x_j}\Big(\frac{\partial f}{\partial x_i}\Big) = \frac{\partial^2 f}{\partial x_j \partial x_i} = \frac{\partial}{\partial x_i}\Big(\frac{\partial f}{\partial x_j}\Big) = \frac{\partial f_j}{\partial x_i}.$$

The converse is true for *suitable* open sets – the whole space $\mathbb{R}^n$ is always suitable and so also are sets defined by linear inequalities, i.e. sets of the form

$$\big\{ X \in \mathbb{R}^n : l_i(X) < c_i \text{ for } i = 1, \ldots, k \big\}$$

where each l_i is a linear mapping from $\mathbb{R}^n$ into $\mathbb{R}$. In $\mathbb{R}^2$ an open set U is suitable if and only if the "interior" of any closed curve in U also lies in U; roughly speaking this means that U contains no holes. In particular the open set $\mathbb{R}^2 \setminus \{(0,0)\}$ *is not* suitable. However, in $\mathbb{R}^3$ the whole space with a finite number of points removed *is* suitable. These examples show that the concept of suitability is rather subtle. The proof of the following result involves Green's theorem (Chapter 9) and Proposition 6.1.

Proposition 6.2

If $F = (f_1, \ldots, f_n)$ is a continuously differentiable vector field on a suitable open set in $\mathbb{R}^n$ then F has a potential if and only if

$$\frac{\partial f_i}{\partial x_j} = \frac{\partial f_j}{\partial x_i} \tag{6.2}$$

for all i and j.

In $\mathbb{R}^3$ condition (6.2) can be rewritten in a symbolic, easily remembered fashion using the *cross product*. Since we need this product in the next chapter we introduce it here. If $\mathbf{v} = (v_1, v_2, v_3)$ and $\mathbf{w} = (w_1, w_2, w_3)$ are two vectors in $\mathbb{R}^3$ then the *cross product* of $\mathbf{v}$ and $\mathbf{w}$, $\mathbf{v} \times \mathbf{w}$, is defined as

$$\mathbf{v} \times \mathbf{w} = \begin{vmatrix} \mathbf{i} & \mathbf{j} & \mathbf{k} \\ v_1 & v_2 & v_3 \\ w_1 & w_2 & w_3 \end{vmatrix}$$

$$= (v_2 w_3 - v_3 w_2)\mathbf{i} - (v_1 w_3 - v_3 w_1)\mathbf{j} + (v_1 w_2 - v_2 w_1)\mathbf{k}$$

$$= (v_2 w_3 - v_3 w_2, -v_1 w_3 + v_3 w_1, v_1 w_2 - v_2 w_1).$$

Since $\nabla f = (\frac{\partial}{\partial x}, \frac{\partial}{\partial y}, \frac{\partial}{\partial z})(f)$ we define curl(F) or $\nabla \times F$, $F = (f_1, f_2, f_3)$ a vector field, by the formula

$$\nabla \times F = \begin{vmatrix} \mathbf{i} & \mathbf{j} & \mathbf{k} \\ \frac{\partial}{\partial x} & \frac{\partial}{\partial y} & \frac{\partial}{\partial z} \\ f_1 & f_2 & f_3 \end{vmatrix}$$

$$= (\frac{\partial f_3}{\partial y} - \frac{\partial f_2}{\partial z})\mathbf{i} - (\frac{\partial f_3}{\partial x} - \frac{\partial f_1}{\partial z})\mathbf{j} + (\frac{\partial f_2}{\partial x} - \frac{\partial f_1}{\partial y})\mathbf{k}.$$

One easily sees that curl$(\nabla f) = 0$ for any real-valued function with first- and second-order partial derivatives while the hypothesis (6.2) says that curl $(F) = 0$. If F is the velocity of a fluid then curl(F) measures the tendency of the fluid to curl or rotate about an axis. Curl(F) gives the direction of the axis of rotation and $\|\text{curl}(F)\|$ measures the speed of rotation.

Example 6.2

We wish to show that $F(x, y, z) = (ye^z, xe^z, xye^z)$ has a potential. Let $F = (f_1, f_2, f_3)$. To apply Proposition 6.1 we must show

$$\frac{\partial f_1}{\partial y} = \frac{\partial f_2}{\partial x}, \quad \frac{\partial f_1}{\partial z} = \frac{\partial f_3}{\partial x}, \quad \frac{\partial f_2}{\partial z} = \frac{\partial f_3}{\partial y}.$$

We have

$$\frac{\partial f_1}{\partial y} = e^z = \frac{\partial f_2}{\partial x}, \quad \frac{\partial f_1}{\partial z} = ye^z = \frac{\partial f_3}{\partial x}, \quad \frac{\partial f_2}{\partial z} = xe^z = \frac{\partial f_3}{\partial z}$$

and hence F has a potential on $\mathbb{R}^3$.

The following is probably the simplest way to find a potential. If f is a potential for F then

$$\frac{\partial f}{\partial x} = ye^z, \quad \frac{\partial f}{\partial y} = xe^z, \quad \frac{\partial f}{\partial z} = xye^z. \tag{6.3}$$

Hence

$$f = \int \frac{\partial f}{\partial x} dx = \int y e^z dx = xye^z + \phi(y,z)$$

where ϕ is the constant of integration *with respect to* x which may, however, depend on y and z. Differentiating with respect to y we get

$$\frac{\partial f}{\partial y} = xe^z + \frac{\partial \phi}{\partial y}$$

and comparing this with the formula for $\dfrac{\partial f}{\partial y}$ in (6.3) we have

$$xe^z + \frac{\partial \phi}{\partial y} = xe^z$$

and

$$\frac{\partial \phi}{\partial y} = 0.$$

Hence ϕ does not depend on y and we let $\phi(y,z) = \psi(z)$. Now differentiating

$$f(x,y,z) = xye^z + \psi(z)$$

and comparing this with (6.3) gives

$$\frac{\partial f}{\partial z} = xye^z + \psi'(z) = xye^z$$

and $\psi'(z) = 0$. This implies that ψ is a constant and we have shown

$$f(x,y,z) = xye^z + c$$

for some constant c.

We can also use Proposition 6.1 to find a potential f. This proposition tells us that

$$f(x,y,z) = \int_\Gamma F$$

where Γ is any piecewise smooth directed curve in $\mathbb{R}^3$, joining a fixed point to (x,y,z), is a potential of F. We take the fixed point to be the origin in $\mathbb{R}^3$ and a piecewise smooth curve Γ consisting of three straight lines parallel to the axis Γ_1, Γ_2, Γ_3. Specifically we use the following:

Γ_1 joins $(0,0,0)$ to $(x,0,0)$, $P_1(t) = (t,0,0)$, $P_1'(t) = (1,0,0)$, $0 \le t \le x$
Γ_2 joins $(x,0,0)$ to $(x,y,0)$, $P_2(t) = (x,t,0)$, $P_2'(t) = (0,1,0)$, $0 \le t \le y$
Γ_3 joins $(x,y,0)$ to (x,y,z), $P_3(t) = (x,y,t)$, $P_3'(t) = (0,0,1)$, $0 \le t \le z$

(see Figure 6.4).

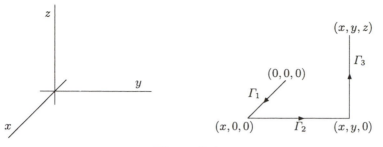

Figure 6.4.

Then

$$f(x, y, z) = \int_{\Gamma_1} F + \int_{\Gamma_2} F + \int_{\Gamma_3} F$$

$$= \int_0^x f_1\big(P_1(t)\big)dt + \int_0^y f_2\big(P_2(t)\big)dt + \int_0^z f_3\big(P_3(t)\big)dt$$

$$= \int_0^x f_1(t,0,0)dt + \int_0^y f_2(x,t,0)dt + \int_0^z f_3(x,y,t)dt$$

$$= \int_0^x 0 \cdot dt + \int_0^y xe^0 dt + \int_0^z xye^t dt$$

$$= 0 + xy + \big[xy \cdot e^t\big]_0^z$$

$$= xy + xye^z - xy$$

$$= xye^z .$$

We verify this result by noting

$$\frac{\partial}{\partial x}(xye^z) = ye^z = f_1, \quad \frac{\partial}{\partial y}(xye^z) = xe^z = f_2, \quad \frac{\partial}{\partial z}(xye^z) = xye^z = f_3.$$

Since the symbolism $\nabla \times F$ has proved useful we consider the analogous symbol $\nabla \cdot F$ where the dot replaces the cross product. This makes sense for a vector field on $\mathbb{R}^n$ and, if $F = (f_1, f_2, \ldots f_n)$, we let

$$\nabla \cdot F = (\frac{\partial}{\partial x_1}, \ldots, \frac{\partial}{\partial x_n}) \cdot (f_1, \ldots, f_n) = \sum_{i=1}^n \frac{\partial f_i}{\partial x_i}.$$

This is called the *divergence* of F and written $\mathrm{div}(F)$ (see Chapter 15). A vector field F on an open subset U of $\mathbb{R}^3$ is said to have a *vector potential* if there exists a vector field G, called a *vector potential*, on U such that

$$\mathrm{curl}\ (G) = \nabla \times G = F.$$

On "suitable" open sets a vector field F has a vector potential if and only if $\nabla \cdot F = \mathrm{div}(F) = 0$.

Example 6.3

We show that $F(X) = X / \|X\|^3$ has a vector potential on $\mathbb{R}^3 \setminus \{ \text{ the } z\text{-axis}\}$. Let

$$G(x,y,z) = \frac{(yz, -xz, 0)}{(x^2 + y^2)(x^2 + y^2 + z^2)^{1/2}} = \frac{z}{\|X\|^2 - z^2} \cdot \frac{1}{\|X\|}(y, -x, 0)$$

on $\mathbb{R}^3 \setminus \{ \text{ the } z\text{-axis}\} = \mathbb{R}^3 \setminus \{(x,y,z) : x = y = 0\}$. We calculate curl (G). To simplify our calculations we use symmetry and the following result which follows immediately from Exercise 1.10:

$$\frac{\partial}{\partial x}(\|X\|) = \frac{x}{\|X\|}.$$

We have

$$\mathrm{curl}\ (G) = (g_1, g_2, g_3) = \begin{vmatrix} \mathbf{i} & \mathbf{j} & \mathbf{k} \\ \dfrac{\partial}{\partial x} & \dfrac{\partial}{\partial y} & \dfrac{\partial}{\partial z} \\ \dfrac{yz}{(x^2 + y^2)\|X\|} & \dfrac{-xz}{(x^2 + y^2)\|X\|} & 0 \end{vmatrix}$$

which implies

$$
\begin{aligned}
g_1 = \frac{\partial}{\partial z}\left(\frac{xz}{(x^2+y^2)\|X\|}\right) &= \frac{x}{x^2+y^2}\frac{\partial}{\partial z}\left(\frac{z}{\|X\|}\right) \\
&= \frac{x}{x^2+y^2} \cdot \frac{\left(\|X\| - \frac{z^2}{\|X\|}\right)}{\|X\|^2} \\
&= \frac{x}{x^2+y^2} \cdot \frac{\left(\|X\|^2 - z^2\right)}{\|X\|^3} \\
&= \frac{x}{\|X\|^3},
\end{aligned}
$$

since $x^2 + y^2 + z^2 = \|X\|^2$. By symmetry $g_2 = y / \|X\|^3$. Finally

$$
\begin{aligned}
g_3 &= \frac{\partial}{\partial x}\left(\frac{-xz}{(x^2+y^2)\|X\|}\right) - \frac{\partial}{\partial y}\left(\frac{yx}{(x^2+y^2)\|X\|}\right) \\
&= -z\left[\frac{(x^2+y^2)\|X\| - 2x^2\|X\| - \frac{x^2(x^2+y^2)}{\|X\|}}{(x^2+y^2)^2\|X\|^2}\right. \\
&\qquad \left. + \frac{(x^2+y^2)\|X\| - 2y^2\|X\| - \frac{y^2(x^2+y^2)}{\|X\|}}{(x^2+y^2)^2\|X\|^2}\right] \\
&= \frac{-z(-1)}{\|X\|^3} = \frac{z}{\|X\|^3}
\end{aligned}
$$

and we have shown

$$\text{curl}\,(G) = \frac{(x,y,z)}{(x^2+y^2+z^2)^{3/2}} = \frac{X}{\|X\|^3} = F.$$

EXERCISES

6.1 Evaluate $\int_\Gamma F$ where F is a vector field and P is a parametrization of the directed curve Γ

(a) $F(x,y,z) = (x,y,z)$, $P(t) = (\sin t, \cos t, t)$, $0 \leq t \leq 2\pi$

(b) $F(x,y,z) = x^2 dx + xyz dy + xz dz$ $P(t) = (t, t^2, t^3)$, $0 \leq t \leq 1$.

(c) $F(x,y,z) = \cos z\mathbf{i} + e^x\mathbf{j} + e^y\mathbf{k}$, $P(t) = (1, t, e^t)$, $0 \leq t \leq 4$.

6.2 Find, if they exist, scalar potentials for the following vector fields

(a) $F(x,y,z) = (xy,\ yz,\ zx)$

(b) $F(x,y,z) = (y,\ z\cos yz,\ y\cos yz)$

(c) $F(x,y,z) = \big(y+yz\cos(xyz),\ x+xz\cos(xyz),\ 2z+xy\cos(xyz)\big)$

(d) $F(x,y,z) = \big(2x\cos(x^2+yz),\ z\cos(x^2+yz),\ y\cos(x^2+yz)\big).$

6.3 Let $f(x,y,z) = x^2y^2 + y^2z^2$. Verify directly that $\nabla \times \nabla f = 0$.

6.4 Compute the curl of each of the following vector fields:

(a) $F_1(x,y,z) = \dfrac{(3,1,2)}{\|X\|}$

(b) $F_2(x,y,z) = \dfrac{(yz,\ zx,\ xy)}{\|X\|}$

(c) $F_3(X) = \dfrac{\langle X,X\rangle X}{\|X\|^4}$

where $X = (x,y,z) \in \mathbb{R}^3 \setminus \{0\}$.

6.5 If $\mathbf{a}$, $\mathbf{b}$, $\mathbf{c}$ and $\mathbf{d}$ are vectors in $\mathbb{R}^3$ show that

(a) $\mathbf{a}\cdot\mathbf{b}\times\mathbf{c} = \mathbf{c}\cdot\mathbf{a}\times\mathbf{b} = \begin{vmatrix} a_1 & a_2 & a_3 \\ b_1 & b_2 & b_3 \\ c_1 & c_2 & c_3 \end{vmatrix}.$

(b) $\mathbf{a}\times(\mathbf{b}\times\mathbf{c}) = (\mathbf{a}\cdot\mathbf{c})\mathbf{b} - (\mathbf{a}\cdot\mathbf{b})\mathbf{c}$

(c) $(\mathbf{a}\times\mathbf{b})\cdot(\mathbf{c}\times\mathbf{d}) = \begin{vmatrix} \mathbf{a}\cdot\mathbf{c} & \mathbf{a}\cdot\mathbf{d} \\ \mathbf{b}\cdot\mathbf{c} & \mathbf{b}\cdot\mathbf{d} \end{vmatrix}.$

6.6 Let $f, g: \mathbb{R}^3 \rightarrow \mathbb{R}$ and $\mathbf{F}, \mathbf{G}: \mathbb{R}^3 \rightarrow \mathbb{R}^3$ denote smooth functions. Prove

(a) $\mathrm{div}(f\mathbf{F}) = f\,\mathrm{div}\mathbf{F} + \nabla f \cdot \mathbf{F}$

(b) $\mathrm{div}(\mathbf{F} \times \mathbf{G}) = \mathrm{curl}(\mathbf{F}) \cdot \mathbf{G} - \mathbf{F} \cdot \mathrm{curl}(\mathbf{G})$

(c) $\mathrm{curl}(f\mathbf{F}) = \nabla f \times \mathbf{F} + f\,\mathrm{curl}(\mathbf{F})$.

6.7 If $f: U$ (open) $\subset \mathbb{R}^3 \rightarrow \mathbb{R}$ has continuous second-order derivatives show that
$$\mathrm{div}(\nabla f) = \frac{\partial^2 f}{\partial x^2} + \frac{\partial^2 f}{\partial y^2} + \frac{\partial^2 f}{\partial z^2}\,.$$

Symbolically the left-hand side has the form $\nabla \cdot \nabla f$ and is written (for this reason) $\nabla^2 f$. ($\nabla^2 f$ is called the *Laplacian* of f and if $\nabla^2 f = 0$ then f is called *harmonic*). Show that
$$\mathrm{curl}\,(\mathrm{curl}\mathbf{F}) = \nabla\,(\mathrm{div}\mathbf{F}) - \nabla^2(\mathbf{F})$$

where $\mathbf{F} = (f_1, f_2, f_3)$ is a vector field and
$$\nabla^2(\mathbf{F}) = (\nabla^2 f_1, \nabla^2 f_2, \nabla^2 f_3)\,.$$

6.8 Show that $\dfrac{z}{(x^2 + y^2 + z^2)^{3/2}}$ is harmonic on $\mathbb{R}^3 \setminus \{0,0,0\}$.

6.9 If $f: \mathbb{R}^+ \rightarrow \mathbb{R}$ and $g(X) = f(\|X\|)$ for $X \in \mathbb{R}^3 \setminus \{0,0,0\}$ show that

$$\nabla g(X) = f'(\|X\|)\frac{X}{\|X\|} \quad \text{and} \quad \nabla^2 g(X) = f''(\|X\|) + \frac{2}{\|X\|}f'(\|X\|)\,.$$

Show that g is harmonic on $\mathbb{R}^3 \setminus \{0,0,0\}$ if and only if
$$f(r) = \frac{A}{r} + B$$

for all $r \neq 0$ in $\mathbb{R}$.

7

The Frenet–Serret Equations

Summary. *We discuss curvature and torsion of directed curves and derive the Frenet–Serret equations. Vector-valued differentiation and orthonormal bases are the main tools used.*

In this chapter we define geometric concepts associated with a directed curve and derive a set of equations – the *Frenet–Serret equations* – which capture the fundamental relationships between them.

We begin with directed curves in $\mathbb{R}^2$ since this particular case exhibits special features not present in higher dimensions. These are due to considering a *one-dimensional* object (the directed curve) in a *two-dimensional* space ($\mathbb{R}^2$). The same phenomena appear in Chapter 12 when we examine a *two-dimensional* object (an oriented surface) in *three-dimensional* space ($\mathbb{R}^3$) and the same underlying principles are present when we introduce *torsion* later in this chapter. Moreover, our motivation and interpretation of normal curvature (Chapter 16) and geodesic curvature (Chapter 18) are based on our study of curves in $\mathbb{R}^2$. This special straightforward case deserves particular attention because of the insight it provides into later developments.

Let $P\colon [a,b] \to \mathbb{R}^2$ denote a *unit speed* parametrization of the directed curve Γ and let $P(t) = \big(x(t), y(t)\big)$ for all t in $[a,b]$. At $P(t) \in \Gamma$ the unit tangent, $T(t)$, is given by

$$T(t) = P'(t) = \big(x'(t), y'(t)\big).$$

The special features, mentioned above, imply that there are just *two* unit vectors in $\mathbb{R}^2$ perpendicular to $T(t)$ and using the *anticlockwise* (or *counterclockwise*) *orientation* of $\mathbb{R}^2$ we can distinguish between them. If we rotate $T(t)$

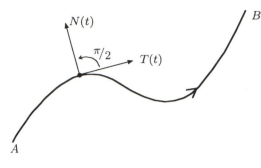

Figure 7.1.

through $+\pi/2$ in an anticlockwise direction we obtain a unit vector on the *left-hand side* of the direction of motion along Γ (Figure 7.1). We call this unit vector the *unit normal* to Γ at $P(t)$ and denote it by $N(t)$. In coordinates

$$N(t) = \left(-y'(t), x'(t)\right).$$

We have $\langle T(t), T(t)\rangle = 1$ and differentiating we get, by the product rule,

$$\frac{d}{dt}\langle T(t), T(t)\rangle = 0 = \langle T'(t), T(t)\rangle + \langle T(t), T'(t)\rangle.$$

Since

$$\langle T'(t), T(t)\rangle = \langle T(t), T'(t)\rangle$$

this implies

$$\langle T'(t), T(t)\rangle = 0.$$

Hence $T'(t)$ is perpendicular to $T(t)$ and, using once more the fact that $\mathbb{R}^2$ is two-dimensional, we see that $T'(t)$ is parallel to $N(t)$. The *curvature* of Γ at $P(t)$ is defined as the unique scalar, $\kappa(t)$, satisfying

$$T'(t) = \kappa(t)N(t). \tag{7.1}$$

In terms of coordinates

$$
\begin{aligned}
\kappa(t) &= \langle \kappa(t)N(t), N(t)\rangle = \langle T'(t), N(t)\rangle \\
&= \left(x''(t), y''(t)\right) \cdot \left(-y'(t), x'(t)\right) \\
&= y''(t)x'(t) - x''(t)y'(t) \tag{7.2}
\end{aligned}
$$

for all $t \in [a, b]$. We call $|\kappa(t)|$ the *absolute curvature* of Γ at $P(t)$ and note that

$$|\kappa(t)| = \|T'(t)\| = \|P''(t)\|. \tag{7.3}$$

Example 7.1

Let $P\colon [a, b] \to \mathbb{R}^2$ denote an *arbitrary* parametrization of the directed curve Γ. We recall from Chapter 5 that $P \circ s^{-1}\colon [0, l] \to \Gamma$ is a unit speed parametrization of Γ where l is the length of Γ and s is the length function. If $P(t) = \big(x(t), y(t)\big)$, $t \in [a, b]$, then

$$P \circ s^{-1}(t) = \big(x \circ s^{-1}(t), y \circ s^{-1}(t)\big), \quad t \in [0, l]$$

and, moreover,

$$(s^{-1})'(t) = \frac{1}{\|P'\big(s^{-1}(t)\big)\|} = \frac{1}{\Big(x'\big(s^{-1}(t)\big)^2 + y'\big(s^{-1}(t)\big)^2\Big)^{1/2}}.$$

We have

$$(x \circ s^{-1})'(t) = x'\big(s^{-1}(t)\big) \cdot (s^{-1})'(t)$$

and

$$(x \circ s^{-1})''(t) = x''\big(s^{-1}(t)\big)\big((s^{-1})'(t)\big)^2 + x'\big(s^{-1}(t)\big) \cdot (s^{-1})''(t)$$

and analogous formulae for $(y \circ s^{-1})'(t)$ and $(y \circ s^{-1})''(t)$. Substituting these into (7.2) and simplifying we obtain the curvature at the point $P(t)$,

$$\kappa(t) = \frac{y''(t)x'(t) - x''(t)y'(t)}{\Big((x'(t))^2 + (y'(t))^2\Big)^{3/2}}. \tag{7.4}$$

If Γ is the graph of a smooth function $f\colon [a, b] \to \mathbb{R}$ directed from left to right then $P(t) = \big(t, f(t)\big)$, $t \in [a, b]$, is a parametrization of Γ. This parametrization is only unit speed in the trivial case of a horizontal line (why?). Since $x(t) = t$ we have $x'(t) = 1$ and $x''(t) = 0$ and as $y(t) = f(t)$ we obtain $y'(t) = f'(t)$ and $y''(t) = f''(t)$. Hence the curvature at $\big(t, f(t)\big)$ is

$$\kappa(t) = \frac{f''(t)}{\Big(1 + f'(t)^2\Big)^{3/2}}.$$

We now discuss the geometric significance of curvature. Let P denote a unit speed parametrization of the directed curve Γ. For simplicity we suppose $0 \in [a, b]$, the domain of definition of P. If t is close to zero then

$$P(t) = P(0) + P'(0)t + P''(0)\frac{t^2}{2} + g(t)t^2 \tag{7.5}$$

where $g(t) \to 0$ as $t \to 0$. Since $P'(0) = T(0)$ and $P''(0) = T'(0) = \kappa(0)N(0)$ we can rewrite this as

$$P(t) = P(0) + T(0)t + \kappa(0)N(0)\frac{t^2}{2} + g(t)t^2.$$

The function

$$Q(t) = P(0) + T(0)t + \kappa(0)N(0)\frac{t^2}{2} \qquad (7.6)$$

is the best *quadratic approximation* to Γ near $P(0)$ and the curve, parametrised by Q, has the *same* tangent, the *same* normal and the *same* curvature as Γ at $P(0)$. By translating and rotating the plane, if necessary, we can suppose $P(0) = (0,0)$, $T(0) = (1,0)$ and $N(0) = (0,1)$. This implies $Q(t) = (t, \kappa(0)t^2/2)$ and, if $\kappa(0) \neq 0$, then one of the two situations portrayed in Figure 7.2 holds.

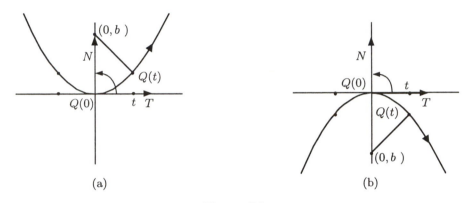

(a)   (b)

Figure 7.2.

Let C_t denote the circle with centre (a_t, b_t) which passes through the points $Q(-t)$, $Q(0)$, $Q(t)$. By symmetry (a_t, b_t) lies on the y-axis, hence $a_t = 0$ and $|b_t|$ is the radius of C_t. As t tends to zero the circles C_t converge to a circle C with centre $(0, b)$ and radius $|b|$. This is the *circle of curvature* at $Q(0)$ to the directed curve parametrised by Q. Since $\frac{P(t)-Q(t)}{t^2} \to 0$ as $t \to 0$ it can easily be shown that C is also the circle of curvature to Γ at $P(0)$, i.e. the circle that fits closest to Γ near $P(0)$. The centre $(0, b)$ is called the *centre of curvature* of Γ at $P(0)$ and $|b|$ is the *radius of curvature*.

We have

$$b_t^2 = \|(0, b_t) - (t, \kappa(0)t^2/2)\|^2 = t^2 + (b_t - \kappa(0)t^2/2)^2.$$

Hence

$$b_t^2 = t^2 + b_t^2 - b_t\kappa(0)t^2 + \kappa(0)^2t^4/4$$

and

$$b_t \kappa(0) = 1 + \frac{\kappa(0)^2 t^2}{4}.$$

Letting t tend to zero we get $b\kappa(0) = 1$. We interpret $|\kappa(0)|$ as

$$\frac{1}{|b|} = \frac{1}{\text{radius of circle of curvature}}.$$

Since the sign of b tells us on which side of Γ the circle of curvature lies and $b\kappa(0) = 1$ we have

$$\kappa(0) > 0 \iff \quad \text{the circle of curvature and the normal are}$$
$$\text{on the } \textit{same} \text{ side of } \Gamma \text{ (Figure 7.2(a))}$$
$$\kappa(0) < 0 \iff \quad \text{the circle of curvature and the normal lie}$$
$$\text{on } \textit{opposite} \text{ sides of } \Gamma \text{ (Figure 7.2(b)).}$$

If $\kappa(0) = 0$ then Γ is rather flat near $P(0)$ and the circle of curvature has infinite radius and thus is a straight line – in our case the x-axis.

We have thus established a geometric interpretation for both absolute curvature and the sign of curvature in $\mathbb{R}^2$ and this often yields immediate and useful information. For example, consider the ellipse in Figure 7.3, oriented in an anticlockwise direction. The normal will always point into the ellipse and is called, for this reason, the *inner normal*. Since the circle of curvature at any point has finite radius and lies on the same side of the tangent line as the curve the curvature is always strictly positive.

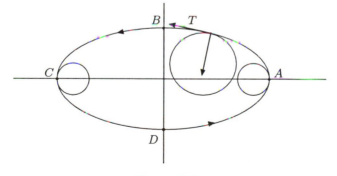

Figure 7.3.

At the points A and C the circles of closest fit to the ellipse have minimum radii among all points on the ellipse. Hence we have maximum curvature at A and C and, similarly, minimum curvature at B and D.

Now consider a directed curve Γ in $\mathbb{R}^n$, $n > 2$, with unit speed para–metrization $P: [a, b] \to \Gamma$. As before the unit tangent to Γ at $P(t)$ is $P'(t) = T(t)$. We cannot, however, define the unit normal to Γ at $P(t)$ by rotating $T(t)$ since there are not just *two* but an *infinite* number of sides to Γ and thus an infinite number of ways of choosing a unit vector perpendicular to $T(t)$. Even if there were only two directions we would require some concept of *anticlockwise* direction in $\mathbb{R}^n$ to distinguish between them.

To define a normal we need to choose a direction perpendicular to T which is associated in some way to the curve Γ. We still have $\langle T(t), T(t) \rangle = 1$ for all t in $[a, b]$ and differentiating, as we did previously, we get $\langle T'(t), T(t) \rangle = 0$. We define the *curvature* $\kappa(t)$ by $\kappa(t) = \|T'(t)\|$.

We have found a special vector, $T'(t)$, perpendicular to $T(t)$ and if this is non-zero (or equivalently if $P''(t) \neq 0$) we define the *unit normal*, $N(t)$, by

$$N(t) = \frac{T'(t)}{\|T'(t)\|} .$$

The normal is *only* defined at points of non-zero curvature and at such points we obtain the equation

$$T'(t) = \kappa(t) N(t) . \tag{7.1$'$}$$

Note that $(7.1)'$ and (7.1) are the *same* equation. However, $(7.1)'$ applies to a curve in $\mathbb{R}^2$ while (7.1), which is the first of the *Frenet–Serret equations* when $n = 3$, applies to a curve in $\mathbb{R}^n$. The definitions of curvature and normal are *different* in these two equations. The technique of using an equation in a simple setting to extend a definition to a more general setting is standard and useful in mathematics.

Now that we have defined κ and N in $\mathbb{R}^n$ we must investigate their properties as they could well be different to those in $\mathbb{R}^2$. Using the same terminology is an expression of our aspirations but does not qualify as a proof. We note first that curvature in $\mathbb{R}^n$ is *always* defined and always non-negative but certain curves, such as straight lines, do not have a normal.

Equation (7.5) and the approximation (7.6) are still valid for curves in $\mathbb{R}^n$ and the same analysis shows that $\kappa(t)$ can be interpreted as the reciprocal of the radius of the circle of curvature to Γ at $P(t)$. Hence curvature in $\mathbb{R}^n$, $n > 2$, has the *same* geometrical interpretation as absolute curvature in $\mathbb{R}^2$. We have seen that the sign of curvature in $\mathbb{R}^2$ was related to the different sides of a curve and, in view of our previous remarks, it is not surprising that it does not feature when $n > 2$.

From now on we restrict ourselves to curves in $\mathbb{R}^3$. The approximation

$$Q(t) = P(0) + T(0)t + \kappa(0)N(0)\frac{t^2}{2}$$

near $0 \in [a, b]$ is still valid and shows that the plane (or two-dimensional sub-space) in $\mathbb{R}^3$ closest to the curve near $P(t)$ is the plane through $P(t)$ spanned by $\{T(t), N(t)\}$. We call this the *osculating plane* of the curve at $P(t)$. We may consider the osculating plane as the two-dimensional analogue of the *tangent line*. As we move along the curve the osculating plane will generally change and the more it changes the more twisted the curve. We measure this by defining a new concept – *torsion* –which we denote by τ. We define torsion and a third unit vector, the *binormal* in $\mathbb{R}^3$, in a fashion similar to the way we introduced curvature and the normal for curves in $\mathbb{R}^2$. At each point on the directed curve Γ in $\mathbb{R}^3$ we have obtained two perpendicular unit vectors T and N and these span a two-dimensional subspace of $\mathbb{R}^3$. Hence there are precisely *two* unit vectors perpendicular to T and N. To choose one of these *unambiguously* we require a sense of *direction* or *orientation* in $\mathbb{R}^3$. This will also be important in integration theory. The basis of this sense of direction is known as *"the right-hand rule"* and we describe it in the special case in which we are interested. Use the right thumb as the vector T and the first finger in place of N. Then the second finger will, when put perpendicular to T and N, give the direction of the *binormal*, B (Figure 7.4).

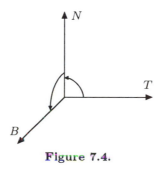

Figure 7.4.

Think of T and N as determining the flat plane of this page. This page has two sides and thus two unit vectors perpendicular to it. Since the vector N is obtained in Figure 7.4 by rotating T in an anticlockwise direction, B will be the unit vector on *this* side of the page. Note that we placed T *before* N in this construction.

Mathematically, we can find the binormal by using the cross product in $\mathbb{R}^3$. For a directed curve Γ in $\mathbb{R}^3$ with parametrization P, unit tangent $T(t)$ and unit normal $N(t)$ the binormal at the point $P(t)$ is given by

$$B(t) = T(t) \times N(t).$$

To derive further results we list standard properties of the *cross product* – all of which follow easily from well-known results about *determinants*.

Let $\mathbf{v}$, $\mathbf{w}$ and $\mathbf{u}$ be vectors in $\mathbb{R}^3$ then

(a) $\mathbf{v} \times \mathbf{w} = -\mathbf{w} \times \mathbf{v}$

(b) $\mathbf{v} \times \mathbf{w} \neq 0 \iff \mathbf{v}$ and $\mathbf{w}$ are linearly independent

(c) $\mathbf{v} \times \mathbf{w}$ is perpendicular to both $\mathbf{v}$ and $\mathbf{w}$

(d) $\|\mathbf{v} \times \mathbf{w}\| = \|\mathbf{v}\| \cdot \|\mathbf{w}\| \cdot |\sin\theta|$, where θ is the angle between $\mathbf{v}$ and $\mathbf{w}$

(e)

$$\mathbf{u} \cdot \mathbf{v} \times \mathbf{w} = \begin{vmatrix} u_1 & u_2 & u_3 \\ v_1 & v_2 & v_3 \\ w_1 & w_2 & w_3 \end{vmatrix}$$

and

$$\mathbf{u} \cdot \mathbf{v} \times \mathbf{w} = \mathbf{w} \cdot \mathbf{u} \times \mathbf{v} = \mathbf{v} \cdot \mathbf{w} \times \mathbf{u}.$$

By (c), $B(t)$ is perpendicular to both $T(t)$ and $N(t)$ and, by (d),

$$\begin{aligned} \|B(t)\| &= \|T(t)\| \cdot \|N(t)\| \cdot |\sin(\pi/2)| \quad (\text{since } T(t) \perp N(t)) \\ &= 1. \end{aligned}$$

Hence $\{T(t), N(t), B(t)\}$ consists of three mutually perpendicular unit vectors in $\mathbb{R}^3$. In particular, they are linearly independent and so form a basis for $\mathbb{R}^3$. We call them an *orthonormal basis* ("ortho" comes from orthogonal or perpendicular and normal comes from the fact that they are unit vectors). Another orthonormal basis for $\mathbb{R}^3$ is the set $\{\mathbf{i}, \mathbf{j}, \mathbf{k}\}$. We consider $\{T(t), N(t), B(t)\}$ as a special basis which is *adapted* to studying the curve Γ near $P(t)$. As t varies the basis $\{T(t), N(t), B(t)\}$ changes and is called a *moving frame* along the curve (see Figure 7.5).

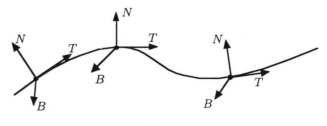

Figure 7.5.

Since $N(t) \times B(t)$ is easily seen to be a unit vector perpendicular to both N and B we must have

$$N(t) \times B(t) = \pm T(t).$$

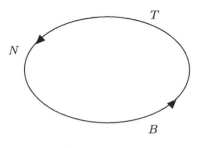

Figure 7.6.

Hence

$$T(t) \cdot N(t) \quad \times \quad B(t) = \pm T(t) \cdot T(t) = \pm 1$$

$$\parallel \quad \text{by (e)}$$

$$B(t) \cdot T(t) \quad \times \quad N(t) = B(t) \cdot B(t) = 1$$

and $N(t) \times B(t) = T(t)$. Similarly $B(t) \times T(t) = N(t)$ and, using (a), the remaining cross products involving T, N and B can be found.

The simplest way to remember these is to use the diagram shown in Figure 7.6.

The cross product of any two taken in an anticlockwise direction is the one that follows it, e.g. $N \times B = T$. If we work in a clockwise direction we get, from (a), the negative of the following one, e.g. $N \times T = -B$.

We are now in a position to make effective use of the orthonormal basis $\{T, N, B\}$. Since it is a basis any vector $\mathbf{v}$ can be written in the form

$$\mathbf{v} = \alpha T + \beta N + \gamma B$$

for some real numbers α, β and γ. If we take the inner product of both sides with respect to T then

$$\langle \mathbf{v}, T \rangle = \alpha \langle T, T \rangle + \beta \langle N, T \rangle + \gamma \langle B, T \rangle = \alpha$$

$$\parallel \qquad\qquad \parallel \qquad\qquad \parallel$$

$$1 \qquad\qquad\quad 0 \qquad\qquad\quad 0$$

Similarly $\beta = \langle \mathbf{v}, N \rangle$ and $\gamma = \langle \mathbf{v}, B \rangle$ and thus

$$\mathbf{v} = \langle \mathbf{v}, T \rangle T + \langle \mathbf{v}, N \rangle N + \langle \mathbf{v}, B \rangle B. \tag{7.7}$$

We also note, although we do not require it here, that Pythagoras' theorem implies

$$\|\mathbf{v}\|^2 = \langle \mathbf{v}, T \rangle^2 + \langle \mathbf{v}, N \rangle^2 + \langle \mathbf{v}, B \rangle^2.$$

Combining (7.2) with vector-valued differentiation we find $B'(t)$. Since $\|B(t)\| = 1$ we have $\langle B(t), B(t) \rangle = 1$ and hence

$$
\begin{aligned}
0 = \frac{d}{dt} \langle B(t), B(t) \rangle &= \langle B'(t), B(t) \rangle + \langle B(t), B'(t) \rangle \\
&= 2 \langle B'(t), B(t) \rangle.
\end{aligned}
\tag{7.8}
$$

Since $\langle B(t), T(t) \rangle = 0$ we get, in the same way,

$$
\begin{aligned}
0 &= \frac{d}{dt} \langle B(t), T(t) \rangle \\
&= \langle B'(t), T(t) \rangle + \langle B(t), T'(t) \rangle \\
&= \langle B'(t), T(t) \rangle + \langle B(t), \kappa(t)N(t) \rangle \quad \text{(by (7.1)}') \\
&= \langle B'(t), T(t) \rangle + \kappa(t) \langle B(t), N(t) \rangle \\
&= \langle B'(t), T(t) \rangle \quad \text{(since } B \perp N).
\end{aligned}
\tag{7.9}
$$

Replacing $\mathbf{v}$ by $B'(t)$ in (7.7) and using (7.8) and (7.9) we have

$$
B'(t) = \langle B'(t), N(t) \rangle N(t)
$$

i.e. $B'(t)$ is parallel to $N(t)$.

We now define the *torsion* of Γ at $P(t)$, $\tau(t)$, by the formula

$$
B'(t) = -\tau(t)N(t).
\tag{7.10}
$$

This is another of the Frenet–Serret equations. We discuss the geometric significance of torsion in the next chapter.

We have found T' and B' in Equations (7.1)' and (7.10) for a directed curve in $\mathbb{R}^3$ and the remaining Frenet–Serret equation is an expression for N' in terms of the basis $\{T, N, B\}$. To find N' we differentiate the equation $N = B \times T$ using the product rule. We have

$$
N' = B' \times T + B \times T' = -\tau N \times T + B \times \kappa N = \tau T \times N - \kappa N \times B.
$$

Since $T \times N = B$ and $N \times B = T$ we obtain for all t the equation

$$
N'(t) = -\kappa(t)T(t) + \tau(t)B(t).
\tag{7.11}
$$

Equations (7.1)', (7.10) and (7.11) which express T', N' and B' in terms of T, N and B are known as the *Frenet–Serret equations* and contain, for all practical purposes, complete information on the curve (see also Example 8.4). The set $\{T, N, B, \kappa, \tau\}$ is known as the *Frenet–Serret apparatus* of the curve Γ.

The Frenet–Serret equations are easily remembered when expressed in matrix form

$$
\begin{pmatrix} T \\ N \\ B \end{pmatrix}' = \begin{pmatrix} 0 & \kappa & 0 \\ -\kappa & 0 & \tau \\ 0 & -\tau & 0 \end{pmatrix} \begin{pmatrix} T \\ N \\ B \end{pmatrix}.
$$

Example 7.2

Let Γ denote the helix parametrised by

$$P(t) = (r\cos\omega t, r\sin\omega t, h\omega t), \quad -\infty < t < +\infty$$

where $\omega = (r^2 + h^2)^{-1/2}$. Since

$$P'(t) = (-r\omega\sin\omega t, r\omega\cos\omega t, h\omega)$$

and

$$
\begin{aligned}
\|P'(t)\| &= (r^2\omega^2\sin^2\omega t + r^2\omega^2\cos^2\omega t + h^2\omega^2)^{1/2} \\
&= (r^2\omega^2 + h^2\omega^2)^{1/2} = \omega(r^2 + h^2)^{1/2} = 1
\end{aligned}
$$

the parametrization is unit speed and $T(t) = P'(t)$. We have

$$P''(t) = T'(t) = (-r\omega^2\cos\omega t, -r\omega^2\sin\omega t, 0)$$

and

$$\kappa(t) = \|T'(t)\| = (r^2\omega^4\cos^2\omega t + r^2\omega^4\sin^2\omega t)^{1/2} = \omega^2 r.$$

Note that Γ is *not* a circle but has constant curvature. Hence $\kappa(t) \neq 0$ and

$$
\begin{aligned}
N(t) &= \frac{T'(t)}{\|T'(t)\|} = \frac{1}{\omega^2 r}(-r\omega^2\cos\omega t, -r\omega^2\sin\omega t, 0) \\
&= (-\cos\omega t, -\sin\omega t, 0).
\end{aligned}
$$

We have

$$
\begin{aligned}
B(t) &= T(t) \times N(t) = \begin{vmatrix} \mathbf{i} & \mathbf{j} & \mathbf{k} \\ -r\omega\sin\omega t & r\omega\cos\omega t & h\omega \\ -\cos\omega t & -\sin\omega t & 0 \end{vmatrix} \\
&= (h\omega\sin\omega t, -h\omega\cos\omega t, r\omega)
\end{aligned}
$$

and

$$B'(t) = (h\omega^2\cos\omega t, h\omega^2\sin\omega t, 0).$$

Since $\tau(t) = -\langle B'(t), N(t)\rangle$ this implies

$$
\begin{aligned}
\tau(t) &= -\big\langle (h\omega^2\cos\omega t, h\omega^2\sin\omega t, 0), (-\cos\omega t, -\sin\omega t, 0)\big\rangle \\
&= h\omega^2\cos^2\omega t + h\omega^2\sin^2\omega t = h\omega^2.
\end{aligned}
$$

We have calculated the Frenet–Serret apparatus for the helix. In doing so we used two of the Frenet–Serret equations. The third equation

$$N' = \kappa T + \tau B$$

may be used to check our calculations. The Frenet–Serret apparatus was found in the following sequence: first check that P is unit speed then

$$\left\{ \begin{array}{ccccc} T(t) & \kappa(t) & N(t) & B(t) & \tau(t) \\ \| & \| & \| & \| & \| \\ P'(t) & \|T'(t)\| & T'(t)/\kappa(t) & T(t) \times N(t) & -\langle B'(t), N(t) \rangle \end{array} \right\}$$

Other sequences are also possible but the above appear to be generally more direct (for unit speed curves).

EXERCISES

7.1 Parametrise the curve $x^2 + (y/3)^2 = 9$ with an anticlockwise orientation and hence find its curvature. Find the points where the curvature is a maximum

   (a) by inspecting a sketch;

   (b) by differentiating the curvature function;

   (c) by inspection of the curvature function.

7.2 Let $f: U$ (open) $\subset \mathbb{R}^2 \to \mathbb{R}$ and suppose $f^{-1}(0)$ has full rank at each point. Show that $f^{-1}(0)$, regarded as a curve in $\mathbb{R}^2$, has absolute curvature
$$\frac{|f_{xx}f_y^2 - 2f_{xy}f_xf_y + f_{yy}f_x^2|}{(f_x^2 + f_y^2)^{3/2}}.$$

Using this result find the curvature at all points on the ellipse $(x/a)^2 + (y/b)^2 = 1$ directed so that the normal points outwards. Verify your answer when $a = 1$ and $b = 3$ using Exercise 7.1.

7.3 Let Γ denote the plane curve parametrised by

$$P(t) = (t, \log \cos t), \quad -\pi/4 \le t \le \pi/4.$$

Show that Γ has curvature $-\cos t$ at $P(t)$.

7.4 Show that a directed curve in $\mathbb{R}^3$ is a straight line if and only if all its tangent lines are parallel.

7.5 Show that each of the following gives a unit speed parametrization of a curve Γ in $\mathbb{R}^3$. Calculate the Frenet–Serret apparatus of the curve and verify that $N' = -\kappa T + \tau B$.

(a) $P(t) = \left(\dfrac{(1+t)^{3/2}}{3}, \dfrac{(1-t)^{3/2}}{3}, \dfrac{t}{\sqrt{2}} \right), \quad 0 \le t \le 1/2$

(b) $P(t) = \dfrac{1}{2}\left(\cos^{-1}(t) - t\sqrt{1-t^2}, 1-t^2, 0 \right), \quad 0 \le t \le 1/2$

(c) $P(t) = \left(\dfrac{(1+t^2)^{1/2}}{\sqrt{5}}, \dfrac{2t}{\sqrt{5}}, \dfrac{\log(t + \sqrt{1+t^2})}{\sqrt{5}} \right), \quad t \in \mathbb{R}.$

7.6 Let $P(t) = (a \cos t, a \sin t, at \tan \alpha)$ denote a parametrization of the helix. Show that the centre of curvature also moves on a helix.

7.7 If a and b are positive real numbers show that the curve parametrised by

$$P(t) = (a \cos t, a \sin t, b \cosh \frac{at}{b}), \quad t \in \mathbb{R}$$

lies on the cylinder $x^2 + y^2 = a^2$. Show that the osculating plane at any point on the curve makes a constant angle with the tangent plane to the cylinder at that point.

7.8 If $P \colon [a, b] \to \Gamma$ is a unit speed parametrised curve show that

$$\langle P' \times P'', P''' \rangle = \kappa^2 \tau$$

and if $\tau \ne 0$ show that

$$\tau = \frac{\langle P' \times P'', P''' \rangle}{\langle P'', P'' \rangle}.$$

7.9 Let P denote a unit speed parametrization of a directed curve in $\mathbb{R}^3$ with non-zero curvature at $P(0)$. Show that the equation of the osculating plane at $P(0)$ is

$$\{ X \in \mathbb{R}^3 : (X - P(0)) \times P'(0) \cdot P''(0) = 0 \}.$$

8
Geometry of Curves in $\mathbb{R}^3$

Summary. *We apply the Frenet–Serret equations to study the geometric signif-icance of torsion, to analyse curves in spheres and to characterise generalised helices.*

We first provide a geometrical interpretation of *zero torsion*.

Proposition 8.1

If Γ is a directed curve in $\mathbb{R}^3$ with positive curvature at all points then the following are equivalent

(a) Γ is a plane curve

(b) the function $t \to B(t)$ is constant

(c) $\tau(t) = 0$ for all t.

Proof

Since $\kappa(t) > 0$, $N(t)$ is defined. By the Frenet–Serret equations $B'(t) = -\tau(t)N(t)$ and since $\|N(t)\| = 1$ we have:

$$B(t) \text{ is independent of } t \iff B'(t) = 0 \iff \tau(t) = 0.$$

Hence (b) and (c) are equivalent. Now suppose Γ is a plane curve, i.e. there exists a plane in $\mathbb{R}^3$ which contains Γ. The results of the previous chapter show that this plane will be the osculating plane at *any* point on the curve. Since the osculating plane is perpendicular to the binormal this proves (a)$\Longrightarrow$(b). We also give an analytic proof of this implication. By rotating $\mathbb{R}^3$ and using a translation, if necessary, we may suppose that Γ lies in the xy–plane, i.e. in $\{(x, y, 0) : x \in \mathbb{R}, y \in \mathbb{R}\}$, and Γ has a parametrization of the form

$$P(t) = \big(x(t), y(t), 0\big), \quad t \in [a, b].$$

Hence $T(t) = \big(x'(t), y'(t), 0\big)$ and $N(t)$ is parallel to $\big(x''(t), y''(t), 0\big)$. The only unit vectors perpendicular to both $T(t)$ and $N(t)$ are $\pm(0, 0, 1)$ and this implies $B(t) = (0, 0, 1)$ or $B(t) = (0, 0, -1)$. Let $f(t) = B(t) \cdot (0, 0, 1)$. The function f is a continuous real-valued function and only takes the values ± 1. If there exist t_1 and t_2 such that $B(t_1) = (0, 0, 1)$ and $B(t_2) = (0, 0, -1)$ then $f(t_1) = 1$ and $f(t_2) = -1$. Since f is continuous on $[a, b]$ the *intermediate value theorem* implies there exists t_0 in (a, b) such that $f(t_0) = 0$. This is impossible. Hence $f(t) = +1$ for all t or $f(t) = -1$ for all t and this implies $B(t) = (0, 0, 1)$ for all t or $B(t) = (0, 0, -1)$ for all t. In either case B is a constant function and (a)$\Longrightarrow$ (b).

Now suppose (b) holds. Let P denote a unit speed parametrization of Γ with domain $[a, b]$. If $t_0 \in (a, b)$, $X_0 = P(t_0)$ and $B(t) = B$ for all t, then

$$\begin{aligned} \frac{d}{dt} \langle P(t) - X_0, B \rangle &= \langle P'(t), B(t) \rangle + \langle P(t) - X_0, 0 \rangle \\ &= \langle T(t), B(t) \rangle + 0 \\ &= 0 \quad \text{(since } T(t) \perp B\text{)}. \end{aligned}$$

Hence there exists a constant c such that

$$\langle P(t) - X_0, B \rangle = c$$

and P lies in the plane through X_0 perpendicular to B. This completes the proof.

Proposition 8.1 gives a precise geometric interpretation of *zero torsion*. To interpret *non-zero torsion* we look at an expansion of the parametrization P about a fixed point relative to the basis $\{T(t_0), N(t_0), B(t_0)\}$. For convenience we may suppose $t_0 = 0$.

In the previous chapter we obtained, using orthogonality, the Taylor series expansion and the Frenet–Serret equations, the following three expansions of $P(t)$:

$$\begin{aligned} P(t) &= \langle P(t), T(0) \rangle T(0) + \langle P(t), N(0) \rangle N(0) + \langle P(t), B(0) \rangle B(0) \quad (8.1) \\ &= P(0) + P'(0)t + P''(0)\frac{t^2}{2} + g(t)t^2 \quad\quad\quad\quad\quad\quad\quad\quad (8.2) \end{aligned}$$

$$= \ P(0) + T(0)t + \frac{\kappa(0)N(0)}{2}t^2 + g(t)t^2 \qquad (8.3)$$

where $g(t) \to 0$ in $\mathbb{R}^3$ as $t \to 0$. From (8.3) we can identify the main influence – the first non-constant term in the Taylor series expansion – on the shape of the curve in the $T(0)$ and $N(0)$ directions. Comparing (8.1), (8.2) and (8.3) we see also that the first possible non-zero term in the $B(0)$ direction will be $\frac{d^3}{dt^3}\langle P(t), B(0) \rangle \Big|_{t=0}$. By repeated differentiation and use of the Frenet–Serret equation for T' we obtain

$$\begin{aligned}
\langle P'(t), B(0) \rangle &= \langle T(t), B(0) \rangle \\
\langle P''(t), B(0) \rangle &= \langle T'(t), B(0) \rangle = \langle \kappa(t)N(t), B(0) \rangle
\end{aligned}$$

and, using the Frenet–Serret equation for N',

$$\begin{aligned}
\langle P'''(t), B(0) \rangle &= \langle \kappa'(t)N(t) + \kappa(t)N'(t), B(0) \rangle \\
&= \langle \kappa'(t)N(t) + \kappa(t)\big(-\kappa(t)T(t) + \tau(t)B(t)\big), B(0) \rangle .
\end{aligned}$$

Letting $t = 0$ we get

$$\begin{aligned}
\langle P'''(0), B(0) \rangle &= \kappa'(0)\langle N(0), B(0) \rangle - \kappa^2(0)\langle T(0), B(0) \rangle \\
&\quad + \kappa(0)\tau(0)\langle B(0), B(0) \rangle \\
&= \kappa(0)\tau(0)
\end{aligned}$$

since $\{T, N, B\}$ are mutually perpendicular unit vectors. This gives us the approximation

$$Q(t) = P(0) + T(0)t + \kappa(0)N(0)\frac{t^2}{2} + \kappa(0)\tau(0)B(0)\frac{t^3}{6}$$

called the *Frenet approximation* to the curve Γ at 0. The Frenet approximation is clearly a refinement of (8.3) which takes account of torsion. The function $t \to Q(t)$ defines a parametrised curve which has the *same* Frenet–Serret apparatus as the original curve at $P(0)$. From the Frenet approximation we see the influence of non-zero torsion on the shape of the curve. Torsion controls the motion of the curve *orthogonal to the osculating plane*. If $\tau(0) > 0$ then the curve twists towards the side of the osculating plane which contains $B(0)$ and the greater $\tau(0)$ the more dramatic the twist. If $\tau(0) < 0$ the curve twists towards $-B(0)$.

An everyday example of non-zero torsion is given by the curve on the edge of a screw. In tightening a screw one usually uses the right-hand and follows the right-hand rule while in loosening a screw one follows the left-hand rule. If you have any doubts about the difference change hands. This also illustrates the two different orientations of $\mathbb{R}^3$.

Example 8.1

In this example we study a directed curve Γ which lies in a sphere with centre c and radius r. Let P denote a unit speed parametrization of Γ. Our hypothesis states that

$$\|P(t) - c\|^2 = \langle P(t) - c, P(t) - c \rangle = r^2.$$

Consider the expansion of $P(t) - c$ relative to the orthonormal basis $\{T(t), N(t), B(t)\}$, i.e.

$$
\begin{aligned}
P(t) - c &= \langle P(t) - c, T(t) \rangle T(t) + \langle P(t) - c, N(t) \rangle N(t) \\
&\quad + \langle P(t) - c, B(t) \rangle B(t).
\end{aligned}
\tag{8.4}
$$

Differentiating we get

$$\frac{d}{dt} \langle P(t) - c, P(t) - c \rangle = 0 = 2 \langle P(t) - c, P'(t) \rangle.$$

Since P is unit speed, $P'(t) = T(t)$, and we may restate this as follows:

$$\langle P(t) - c, T(t) \rangle = 0. \tag{8.5}$$

Differentiating again and using the Frenet–Serret equation for T' gives us

$$
\begin{aligned}
0 = \frac{d}{dt} \langle P(t) - c, T(t) \rangle &= \langle T(t), T(t) \rangle + \langle P(t) - c, T'(t) \rangle \\
&= 1 + \kappa(t) \langle P(t) - c, N(t) \rangle.
\end{aligned}
$$

Hence $\kappa(t) \neq 0$ for all t, $N(t)$ is defined and

$$\langle P(t) - c, N(t) \rangle = -\frac{1}{\kappa(t)}. \tag{8.6}$$

Differentiating (8.6) and applying the Frenet–Serret equation for N' we obtain

$$
\begin{aligned}
-\left(\frac{1}{\kappa(t)}\right)' &= \frac{d}{dt} \langle P(t) - c, N(t) \rangle \\
&= \langle T(t), N(t) \rangle + \langle P(t) - c, N'(t) \rangle \\
&= \langle P(t) - c, -\kappa(t)T(t) + \tau(t)B(t) \rangle \quad (\text{since } N \perp T) \\
&= \tau(t) \langle P(t) - c, B(t) \rangle \quad (\text{by } (8.5)).
\end{aligned}
$$

If $\tau(t) \neq 0$ for all t then

$$\langle P(t) - c, B(t) \rangle = -\frac{1}{\tau(t)} \left(\frac{1}{\kappa(t)}\right)'. \tag{8.7}$$

Substituting (8.5), (8.6) and (8.7) into (8.4) we get

$$P(t) - c = -\frac{1}{\kappa(t)} N(t) - \frac{1}{\tau(t)} \left(\frac{1}{\kappa(t)}\right)' B(t)$$

and we have found $P(t)$ in terms of the Frenet–Serret apparatus of Γ. By Pythagoras' theorem

$$r^2 = \|P(t) - c\|^2 = \frac{1}{\kappa(t)^2} + \left(\left(\frac{1}{\kappa(t)} \right)' \cdot \frac{1}{\tau(t)} \right)^2 .$$

Hence we have recovered the radius of the sphere from the curvature and torsion *when we know* that the curve lies in a sphere. In particular, we see that $r^2 \geq 1/\kappa(t)^2$, or $\kappa(t) \geq 1/r$, which we may loosely rephrase as saying that a curve in a sphere is at least as curved as the sphere in which it lies (our proof only applies to curves of non-zero torsion but it is a simple exercise to show that this is always the case).

Example 8.2

The helix in Example 7.2 satisfies $\langle T_p, (0,0,1) \rangle = h\omega$ for all p where T_p is the unit tangent at the point p. We generalise this by defining a *generalised helix* in $\mathbb{R}^3$, as a curve Γ of positive curvature (at all points) for which there exists a unit vector $\mathbf{u}$ in $\mathbb{R}^3$ such that

$$\langle T_p, \mathbf{u} \rangle = c \quad \text{(constant)}$$

for all $p \in \Gamma$. Let P denote a unit speed parametrization of the generalised helix Γ and let $\{T(t), N(t), B(t), \kappa(t), \tau(t)\}$ denote the Frenet–Serret apparatus at $P(t)$. We prove the following characterisation:

$$\Gamma \text{ is a generalised helix} \iff \frac{\tau(t)}{\kappa(t)} \text{ is constant (i.e. independent of } t).$$

Since $\langle T(t), \mathbf{u} \rangle$ is constant it follows, by the Cauchy–Schwarz inequality (Example 3.4), that

$$|\langle T(t), \mathbf{u} \rangle| \leq \|T(t)\| \cdot \|\mathbf{u}\| \leq 1$$

and $\langle T(t), \mathbf{u} \rangle = \cos \theta$ for some θ. If $\theta = n\pi$ then, by the equality case in the Cauchy–Schwarz inequality, $T(t) = \pm\mathbf{u}$. By the intermediate value theorem (see the proof of Proposition 8.1) this implies that $T(t)$ is a constant function of t. Hence $T'(t) = 0$ and this contradicts our hypothesis. We thus have $\langle T(t), \mathbf{u} \rangle = \cos \theta$ for some $\theta \neq n\pi$. Again using the orthonormal basis $\{T(t), N(t), B(t)\}$ we have

$$\mathbf{u} = \langle \mathbf{u}, T(t) \rangle T(t) + \langle \mathbf{u}, N(t) \rangle N(t) + \langle \mathbf{u}, B(t) \rangle B(t).$$

Since $\langle T(t), \mathbf{u} \rangle = \cos \theta$

$$\frac{d}{dt} \left(\langle T(t), \mathbf{u} \rangle \right) = 0 = \langle T'(t), \mathbf{u} \rangle + \langle T(t), (\mathbf{u})' \rangle = \langle \kappa(t) N(t), \mathbf{u} \rangle$$

(since $\mathbf{u}$ is a constant, $(\mathbf{u})' = 0$). By our hypothesis $\kappa(t) \neq 0$ and this implies $\langle \mathbf{u}, N(t) \rangle = 0$. Hence

$$\mathbf{u} = \cos\theta T(t) + \sin\theta B(t).$$

A further application of the Frenet–Serret equations implies

$$
\begin{aligned}
0 = \frac{d}{dt}\mathbf{u} &= \cos\theta T'(t) + \sin\theta B'(t) \\
&= \cos\theta\kappa(t)N(t) - \sin\theta\tau(t)N(t) \\
&= \big(\cos\theta\kappa(t) - \sin\theta\tau(t)\big)N(t).
\end{aligned}
$$

Since $\|N(t)\| = 1$ this implies $\cos\theta\kappa(t) = \sin\theta\tau(t)$ and

$$\frac{\tau(t)}{\kappa(t)} = \frac{\cos\theta}{\sin\theta} = \cot\theta.$$

We have shown $\tau(t)/\kappa(t)$ is constant for any generalised helix. In obtaining this result we obtained a formula for $\mathbf{u}$ and now use this to prove the converse. Let Γ denote a directed curve with non-zero curvature in $\mathbb{R}^3$ such that $\tau(t)/\kappa(t)$ is constant (i.e. independent of t). This hypothesis implies that there exists a real number θ, $0 < \theta < \pi$, such that $\cot\theta = \tau(t)/\kappa(t)$ for all t (note that θ does not depend on t). Let

$$\mathbf{u}(t) = \cos\theta\, T(t) + \sin\theta\, B(t).$$

By Pythagoras' theorem $\|\mathbf{u}\|^2 = \cos^2\theta + \sin^2\theta = 1$ and $\mathbf{u}$ is a unit vector. To show that $\mathbf{u}$ does not depend on t we prove $\dfrac{d}{dt}\big(\mathbf{u}(t)\big) = 0$. By the Frenet–Serret equations

$$
\begin{aligned}
\frac{d}{dt}\big(\mathbf{u}(t)\big) &= \cos\theta\, T'(t) + \sin\theta\, B'(t) \\
&= \big(\kappa(t)\cos\theta - \tau(t)\sin\theta\big)N(t) = 0.
\end{aligned}
$$

Hence $\mathbf{u}$ does not depend on t and so is a constant. Moreover, since $T \perp B$,

$$\langle T(t), \mathbf{u} \rangle = \langle T(t), \cos\theta\, T(t) + \sin\theta\, B(t) \rangle = \cos\theta.$$

This shows that Γ is a generalised helix and justifies our claim.

Our analysis so far applies to unit speed parametrizations of a directed curve. Unfortunately, many natural parametrizations of curves are not unit speed. It is thus useful to be able to calculate the Frenet–Serret apparatus directly from an arbitrary parametrization.

Let $P\colon [a, b] \to \mathbb{R}^3$ denote an *arbitrary* parametrization of the directed curve Γ. We suppose $P'(t)$ and $P''(t)$ are both non-zero for all t. Let $s\colon [a, b] \to [0, l]$

denote the length function associated with P (see Chapter 5). Then l is the length of Γ, $\|P'(t)\| = s'(t)$ and $Q := P \circ s^{-1}$ is a unit speed parametrization of Γ. Let $\{T(t), N(t), B(t), \kappa(t), \tau(t)\}$ denote the Frenet–Serret apparatus *at the point $P(t)$* on Γ. We have

$$Q \circ s(t) = P \circ s^{-1} \circ s(t) = P(t).$$

Hence

$$\frac{d}{dt}\Big(Q\big(s(t)\big)\Big) = Q'\big(s(t)\big)s'(t) = P'(t) = \|P'(t)\|T(t)$$

and

$$Q'\big(s(t)\big) = T(t) = \frac{P'(t)}{\|P'(t)\|}. \tag{8.8}$$

Differentiating again

$$
\begin{aligned}
\frac{d^2}{dt^2}\Big(Q\big(s(t)\big)\Big) &= \frac{d}{dt}\Big(Q'\big(s(t)\big) \cdot s'(t)\Big) \\
&= Q''\big(s(t)\big)\big(s'(t)\big)^2 + Q'\big(s(t)\big)s''(t) = P''(t).
\end{aligned}
$$

Since Q has unit speed the Frenet–Serret equations imply

$$Q''\big(s(t)\big) = \kappa(t)N(t)$$

and

$$P''(t) = \big(s'(t)\big)^2\kappa(t)N(t) + s''(t)T(t).$$

Hence

$$
\begin{aligned}
P'(t) \times P''(t) &= s'(t)T(t) \times \Big(s'(t)^2\kappa(t)N(t) + s''(t)T(t)\Big) \\
&= s'(t)^3\kappa(t)B(t)
\end{aligned}
$$

since $T \times N = B$ and $T \times T = 0$. Since $\|B(t)\| = 1$ and $s'(t) = \|P'(t)\|$ this implies

$$\kappa(t) = \frac{\|P'(t) \times P''(t)\|}{\|P'(t)\|^3} \tag{8.9}$$

and

$$B(t) = \frac{P'(t) \times P''(t)}{\|P'(t) \times P''(t)\|}. \tag{8.10}$$

The simplest way to obtain N is to use the formula

$$N = B \times T. \tag{8.11}$$

The Frenet–Serret equation for N' implies

$$
\begin{aligned}
Q'''\big(s(t)\big) = (\kappa N)'(t) &= \kappa'(t)N(t) + \kappa(t)N'(t) \\
&= \kappa'(t)N(t) + \kappa(t)\big(-\kappa(t)T(t) + \tau(t)B(t)\big)
\end{aligned}
$$

and hence

$$\langle Q'''(s(t)), B(t)\rangle = \kappa(t)\tau(t).$$

On the other hand

$$
\begin{aligned}
\frac{d^3}{dt^3}\big(Q(s(t))\big) &= P'''(t) \\
&= Q'''(s(t))s'(t)^3 + 3Q''(s(t))s'(t)s''(t) + Q'(s(t))s'''(t) \\
&= Q'''(s(t))s'(t)^3 + 3\kappa(t)s'(t)s''(t)N(t) + s'''(t)T(t).
\end{aligned}
$$

By orthogonality

$$
\begin{aligned}
\left\langle P'''(t), \frac{P'(t) \times P''(t)}{\|P'(t) \times P''(t)\|}\right\rangle &= \langle P'''(t), B(t)\rangle = s'(t)^3\langle Q'''(s(t)), B(t)\rangle \\
&= \|P'(t)\|^3\kappa(t)\tau(t).
\end{aligned}
$$

Finally

$$\tau(t) = \frac{\langle P'''(t), P'(t) \times P''(t)\rangle}{\|P'(t) \times P''(t)\| \cdot \|P'(t)\|^3\kappa(t)} = \frac{\langle P'''(t), P'(t) \times P''(t)\rangle}{\|P'(t) \times P''(t)\|^2}. \tag{8.12}$$

Equations (8.8), (8.9), (8.10), (8.11) and (8.12) are the Frenet–Serret apparatus at $P(t)$ for Γ in terms of the parametrization P.

Example 8.3

Calculate the Frenet–Serret apparatus of the curve parametrised by

$$P(t) = (t - \cos t, \sin t, t).$$

We first calculate the required derivatives of P; P', P'' and P'''. We have $P'(t) = (1 + \sin t, \cos t, 1)$, $P''(t) = (\cos t, -\sin t, 0)$ and $P'''(t) = (-\sin t, -\cos t, 0)$. Next, we obtain the cross product

$$P'(t) \times P''(t) = \begin{vmatrix} \mathbf{i} & \mathbf{j} & \mathbf{k} \\ 1 + \sin t & \cos t & 1 \\ \cos t & -\sin t & 0 \end{vmatrix} = (\sin t, \cos t, -\sin t - 1)$$

and finally the norms or lengths

$$
\begin{aligned}
\|P'(t)\| &= (1 + 2\sin t + \sin^2 t + \cos^2 t + 1)^{1/2} = (3 + 2\sin t)^{1/2} \\
\|P'(t) \times P''(t)\| &= (\sin^2 t + \cos^2 t + \sin^2 t + 2\sin t + 1)^{1/2} \\
&= (2 + 2\sin t + \sin^2 t)^{1/2}.
\end{aligned}
$$

Hence

$$T(t) = \frac{P'(t)}{\|P'(t)\|} = \frac{(1 + \sin t, \cos t, 1)}{(3 + 2\sin t)^{1/2}}$$

$$\kappa(t) = \frac{\|P'(t) \times P''(t)\|}{\|P'(t)\|^3} = \frac{(2 + 2\sin t + \sin^2 t)^{1/2}}{(3 + 2\sin t)^{3/2}}$$

$$B(t) = \frac{P'(t) \times P''(t)}{\|P'(t) \times P''(t)\|} = \frac{(\sin t, \cos t, -\sin t - 1)}{(3 + 2\sin t + \sin^2 t)^{1/2}}$$

$$\tau(t) = \frac{\langle P'''(t), P'(t) \times P''(t) \rangle}{\|P'(t) \times P''(t)\|^2}$$

$$= \frac{(-\sin t, -\cos t, 0) \cdot (\sin t, \cos t, -\sin t - 1)}{2 + 2\sin t + \sin^2 t}$$

$$= \frac{-1}{2 + 2\sin t + \sin^2 t}$$

$$N(t) = B(t) \times T(t)$$

$$= \frac{1}{\sqrt{3 + 2\sin t}\sqrt{2 + 2\sin t + \sin^2 t}} \begin{vmatrix} \mathbf{i} & \mathbf{j} & \mathbf{k} \\ \sin t & \cos t & -\sin t - 1 \\ 1 + \sin t & \cos t & 1 \end{vmatrix}$$

$$= \frac{(2\cos t + \sin t \cos t, -1 - 3\sin t - \sin^2 t, -\cos t)}{(6 + 10\sin t + 7\sin^2 t + 2\sin^3 t)^{1/2}}$$

Example 8.4

In Example 8.1 we showed that curvature and torsion together allowed us to deduce properties of spherical curves. In this example we show that curvature and torsion completely determine the shape of a curve in $\mathbb{R}^3$. Let Γ and Γ_1 denote two directed curves having the same length l in $\mathbb{R}^3$. We suppose that both have positive curvature. Now transfer Γ_1 so that its initial point coincides with the initial point of Γ and rotate it so that the tangents, normals and binormals of Γ and Γ_1 coincide at the initial point. These operations do not affect the shape of Γ_1. Let $P: [0, l] \to \Gamma$ and $P_1: [0, l] \to \Gamma_1$ denote unit speed parametrizations. We now suppose that the curvature and torsion of Γ and Γ_1 at $P(t)$ and $P_1(t)$ coincide for all t and thus we have the Frenet–Serret apparatus $\{T, N, B, \kappa, \tau\}$ and $\{T_1, N_1, B_1, \kappa, \tau\}$ for Γ and Γ_1 respectively. Using the dot product in $\mathbb{R}^3$ we define $g: [0, l] \to \mathbb{R}$ by

$$g(t) = T(t) \cdot T_1(t) + N(t) \cdot N_1(t) + B(t) \cdot B_1(t).$$

By our hypothesis $g(0) = 3$ and by the Cauchy–Schwarz inequality (Example 3.4) $-3 \le g(t) \le 3$ for all t and $g(t) = 3$ if and only if $T(t) = T_1(t)$, $N(t) = N_1(t)$ and $B(t) = B_1(t)$. From the Frenet–Serret equations

$$g' = \kappa N \cdot T_1 + \kappa T \cdot N_1 + (-\kappa T + \tau B) \cdot N_1 + N \cdot (-\kappa T_1 + \tau B_1) - \tau N \cdot B_1 - \tau B \cdot N_1 = 0.$$

Hence g is a constant mapping and, since $g(0) = 3$, we have $g(t) = 3$ for all t

and the Frenet–Serret apparatus is the same for both curves. In particular

$$(P - P_1)'(t) = P'(t) - P_1'(t) = T(t) - T_1(t) = 0$$

and $P(t) = P_1(t) + C$ for all t. Since $P(0) = P_1(0)$ this implies $P(t) = P_1(t)$ for all t and one curve lies on top of the other. We conclude that Γ and Γ_1 have the same shape.

EXERCISES

8.1 Show that a directed curve in $\mathbb{R}^3$ with zero torsion is a generalised helix and hence, or otherwise, give an example of a generalised helix which is not a helix (in the classical sense of Example 7.2).

8.2 Let $P: [0, l] \to \mathbb{R}^3$ denote a unit speed parametrization of a directed curve Γ, with positive curvature and torsion. For $t \in [0, l]$ let

$$Q(t) = \int_0^t B(x)dx, \quad 0 \le t \le l.$$

Show that Q defines a unit speed parametrization of a directed curve $\widetilde{\Gamma}$. If $\{T, N, B, \kappa, \tau\}$ is the Frenet–Serret apparatus for Γ show that $\{B, -N, T, \tau, \kappa\}$ is the Frenet–Serret apparatus for $\widetilde{\Gamma}$.

8.3 Let $P(t) = (t, 1+t^{-1}, t^{-1}-t), \quad 1 \le t \le 2$, denote a parametrization of the curve Γ in $\mathbb{R}^3$. Show that $B(t) = (1/\sqrt{3}, -1/\sqrt{3}, 1/\sqrt{3})$ for all t and hence deduce that Γ lies in the plane $x - y + z = -1$. Find the Frenet–Serret apparatus for Γ.

8.4 If Γ is parametrised by

$$P(\theta) = (\log\cos\theta, \log\sin\theta, \sqrt{2}\theta), \quad \frac{\pi}{4} \le \theta \le \frac{\pi}{3}$$

show that Γ has curvature $\sin 2\theta / \sqrt{2}$ at $P(\theta)$.

8.5 For the curve parametrised by $P(t) = (3t^2, 3t - t^3, 3t + t^3), -1 \le t \le 1$, show that

$$\kappa(t) = -\tau(t) = \frac{1}{3(1 + t^2)^2}.$$

Find a unit vector $\mathbf{u}$ such that $\langle T(t), \mathbf{u} \rangle$ is independent of t.

8.6 Find the curvature and torsion of the curve parametrised by

$$P(t) = (e^t \cos t, e^t \sin t, e^t), \quad t \in \mathbb{R}.$$

8.7 The plane through a point on a curve perpendicular to the tangent line is called the *normal plane* at the point. Show that a curve lies on a sphere if the intersection of all normal planes is non-empty. Hence show that the curve parametrised by

$$P(\theta) = (-\cos 2\theta, -2\cos\theta, \sin 2\theta), \quad \theta \in [0, 2\pi]$$

lies in a sphere. Find the centre and radius of the sphere.

8.8 Show that the curve parametrised by $P(t) = (at, bt^2, t^3)$ is a generalised helix if and only if $4b^4 = 9a^2$.

8.9 If $T: \mathbb{R}^n \to \mathbb{R}^n$ is a linear mapping such that $\|TX\| = \|X\|$ for all $X \in \mathbb{R}^n$ show that T preserves the inner product, angles, area and the length of curves. When $n = 3$, show that T preserves the cross product.

9
Double Integration

Summary. *We define the double integral of a function over an open subset of* $\mathbb{R}^2$ *and use Fubini's theorem to evaluate such integrals. We discuss the fundamental theorem of calculus in* $\mathbb{R}^2$ *– Green's theorem.*

We discuss *(double)* integration of a real-valued function of *two* variables $f(x, y)$ over an open set Ω in $\mathbb{R}^2$. Motivated by the one-dimensional theory we divide Ω into rectangles – the natural analogue of intervals – by first drawing horizontal and vertical lines and thus partitioning the x- and y-axes.

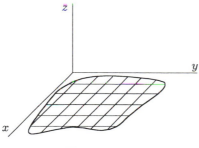

Figure 9.1.

Let x_i denote a typical element of the partition of the the x-axis and let y_j be a typical element on the y-axis. The resulting grid of rectangles gives a partition of Ω (Figure 9.1) and we form the Riemann sum

$$\sum_i \sum_j f(x_i, y_j) \cdot \Delta x_i \cdot \Delta y_j, \quad \Delta x_i = x_{i+1} - x_i, \quad \Delta y_j = y_{j+1} - y_j$$

where we sum over all rectangles which are strictly contained in Ω. If this sum tends to a limit as we take finer and finer partitions we say that f is *integrable over* Ω and denote the limit by

$$\iint\limits_{\Omega} f(x,y)\, dx dy.$$

We call this the *integral* (or *double integral*) of f over Ω. If Ω is the inside of a closed curve Γ and f is *continuous* on $\overline{\Omega}$ it can be shown that f is integrable over Ω. When $f(x,y) = 1$ for all $(x,y) \in \Omega$ the Riemann sum is the area of the rectangles in the partition inside Γ and on taking a limit we obtain

$$\iint\limits_{\Omega} dx dy = \text{Area of } \Omega.$$

If $f(x,y) \geq 0$ then the *volume* of the solid over Ω and beneath the graph of f is

$$\iint\limits_{\Omega} f(x,y) dx dy .$$

We only evaluate double integrals over rather simple open sets. An open set is said to be of *type I* if it is bounded above by the graph of a continuous function $y = h(x)$, bounded below by the graph of a continuous function $y = g(x)$ and on the left and right by vertical lines of finite length (see Figure 9.2).

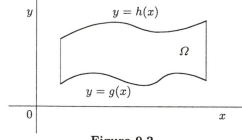

Figure 9.2.

Take a fixed interval in the partition of the x-axis, say (x_i, x_{i+1}), and consider the terms in the Riemann sum

$$\sum_i \sum_j f(x_i, y_j) \cdot \Delta x_i \cdot \Delta y_j$$

which only involve $\Delta x_i = x_{i+1} - x_i$. This gives the sum

$$\left(\sum_j f(x_i, y_j) \cdot \Delta y_j \right) \Delta x_i.$$

Taking limits – this can be justified for *continuous* functions – we get

$$\sum_j f(x_i, y_j) \cdot \Delta y_j \longrightarrow \int_{g(x_i)}^{h(x_i)} f(x_i, y) \, dy$$

as we take finer and finer partitions of the y-axis. Let

$$H(x) = \int_{g(x)}^{h(x)} f(x, y) \, dy.$$

Then

$$\sum_{i,j} f(x_i, y_j) \cdot \Delta x_i \cdot \Delta y_j \approx \sum_i H(x_i) \cdot \Delta x_i \rightarrow \int_a^b H(x) \, dx$$

and taking the limit on both sides we get

$$\iint_\Omega f(x, y) \, dy dx = \int_a^b H(x) \, dx = \int_a^b \left\{ \int_{g(x)}^{h(x)} f(x, y) \, dy \right\} dx.$$

This method of integration, together with the similar method obtained by reversing the roles of x and y, is known as *Fubini's theorem*. We define an open set to be of *type II* if it is bounded on the left and right by the graphs of continuous functions of y, k and l, which are defined on the interval $[c, d]$ and above and below by horizontal lines of finite length. For domains of type II Fubini's theorem is

$$\iint_\Omega f(x, y) \, dx dy = \int_c^d \left\{ \int_{k(y)}^{l(y)} f(x, y) \, dx \right\} dy.$$

If we are given an open set we can recognise that it is of type I if each *vertical* line cuts it at two points except possibly at the end points and type II if each *horizontal* line cuts it at two points except, perhaps, at the end points.

Example 9.1

Evaluate

$$\iint_\Omega \frac{x}{\sqrt{16 + y^7}} \, dx dy$$

over the set Ω bounded above by the line $y = 2$, below by the graph of $y = x^{1/3}$
and on the left by the y-axis (Figure 9.3). By inspection the domain Ω is of
type I and type II and we have a choice of method, i.e. we can integrate first
with respect to either variable. Our choice may be important since one method
may be very simple and the other quite difficult.

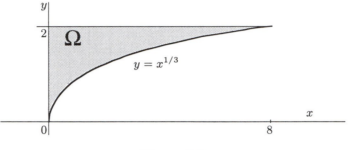

Figure 9.3.

We have to evaluate two integrals of a single variable. In the first integral, the inner integral, one of the variables takes a fixed value and is really a
constant. Thus we have to first evaluate either

$$\int x\,dx \qquad \text{or} \qquad \int \frac{dy}{\sqrt{16 + y^7}}.$$

In these situations looks are usually not deceiving and we opt for the simpler
looking integral. So, we choose to integrate first with respect to x. The limits of
integration in the first integral will influence the degree of difficulty that arises
in evaluating the second, or outer, integral. If you run into problems with the
second integral you should consider starting again using a different order of
integration. In our case we have decided to consider

$$\int \left\{ \int \frac{x}{\sqrt{16 + y^7}}\,dx \right\} dy$$

and now need to determine the variation in x for fixed y. We draw a typical
line through Ω on which y is constant – i.e. a horizontal line.

We must now express the end points in terms of y. Using once more Figure 9.3 we see that the end points of the line of variation of x are $(0, y)$ and
(x, y) where $y = x^{1/3}$. Hence $y^3 = x$ and we have the required variation of x
(Figure 9.4).

Figure 9.4.

We see also that y varies from 0 to 2. Hence

$$\iint_{\Omega} \frac{x}{\sqrt{16 + y^7}} \, dx dy = \int_0^2 \left\{ \int_0^{y^3} \frac{x \, dx}{\sqrt{16 + y^7}} \right\} dy$$

$$= \int_0^2 \frac{x^2 \, dx}{2\sqrt{16 + y^7}} \bigg|_0^{y^3} dy$$

$$= \int_0^2 \frac{y^6}{2\sqrt{16 + y^7}} \, dy$$

$$= \frac{1}{14} \int \frac{dw}{\sqrt{w}} \qquad \begin{aligned} w &= 16 + y^7 \\ dw &= 7y^6 \, dy \end{aligned}$$

$$= \frac{1}{14} \cdot \frac{w^{1/2}}{1/2} = \frac{1}{7}(16 + y^7)^{1/2} \bigg|_0^2$$

$$= \frac{1}{7}\left((144)^{1/2} - (16)^{1/2}\right) = \frac{8}{7}.$$

Example 9.2

In this example we reverse the order of integration and evaluate

$$\int_0^3 \left\{ \int_1^{\sqrt{4-y}} (x + y) \, dx \right\} dy.$$

From the limits of integration in the inner integral the left-hand side of the domain of integration, Ω, is bounded by the line $x = 1$ and the right-hand side by points satisfying $x = \sqrt{4 - y}$, i.e. $x^2 = 4 - y$ or $y = 4 - x^2$. Hence the right-hand side is bounded by the graph of $y = 4 - x^2$ and we have the following diagram for our domain (Figure 9.5).

Reversing the order of integration we get

$$\int_1^2 \left\{ \int_0^{4-x^2} (x + y) \, dy \right\} dx = \int_1^2 \left(xy + \frac{y^2}{2} \right) \bigg|_0^{4-x^2} dx$$

$$= \int_1^2 \left(x(4 - x^2) + \frac{(4 - x^2)^2}{2} \right) dx = \frac{241}{60} \quad \text{(eventually)}.$$

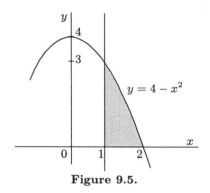

Figure 9.5.

The *fundamental theorem of one-variable calculus*

$$f(b) - f(a) = \int_a^b f'(t)dt \tag{9.1}$$

is used to evaluate integrals of certain functions over intervals from their boundary values. In this theorem integration on the right-hand side is over a *directed interval* while on the left positive and negative signs are assigned to the initial and final points of the interval respectively. Thus we see that a certain coherence has to be established between the orientations on the two sides of (9.1). The *fundamental theorem of two-variable calculus* is known as *Green's theorem*. To obtain this result by an immediate application of the one-variable theorem it is usual to begin with an open subset Ω in $\mathbb{R}^2$ which is of type I and type II. In applying (9.1) it is necessary to be careful with signs and this means that in Green's theorem the boundary of Ω, Γ, is oriented in an *anticlockwise* or *counterclockwise* direction.

Theorem 9.1 (Green's Theorem)

Let P and Q denote real-valued functions with continuous first-order partial derivatives on the open subset U of $\mathbb{R}^2$. If Γ is a closed curve directed in an anticlockwise direction such that the interior (inside) Ω of Γ is an open set of type I and type II and $\Gamma \cup \Omega \subset U$ then

$$\int_\Gamma Pdx + Qdy = \iint_\Omega (\frac{\partial Q}{\partial x} - \frac{\partial P}{\partial y})dxdy. \tag{9.2}$$

Proof

We show

$$\int_\Gamma Q\,dy = \iint_\Omega \frac{\partial Q}{\partial y}\,dx\,dy$$

and for this we use the type II property of Ω. From the representation in Figure 9.6 we have

$$\iint_\Omega \frac{\partial Q}{\partial x}\,dx\,dy = \int_c^d \Big\{\int_{k(y)}^{l(y)} \frac{\partial Q}{\partial x}\,dx\Big\}\,dy.$$

By (9.1)

$$\int_{k(y)}^{l(y)} \frac{\partial Q}{\partial x}(x,y)\,dx = Q(x,y)\,\Big|_{k(y)}^{l(y)} = Q\big(l(y),y\big) - Q\big(k(y),y\big)$$

and

$$\iint_\Omega \frac{\partial Q}{\partial x}\,dx\,dy = \int_c^d \Big(Q\big(l(y),y\big) - Q\big(k(y),y\big)\Big)\,dy.$$

On the other hand, using the parametrizations $y \to \big(l(y),y\big)$ and $y \to \big(k(y),y\big)$, we obtain

$$\begin{aligned}
\int_\Gamma Q\,dy &= \int_{(\text{graph of } l)} Q\,dy - \int_{(\text{graph of } k)} Q\,dy \\
&= \int_c^d Q\big(l(y),y\big)\,dy - \int_c^d Q\big(k(y),y\big)\,dy.
\end{aligned}$$

This proves

$$\iint_\Omega \frac{\partial Q}{\partial x}\,dx\,dy = \int_\Gamma Q\,dy.$$

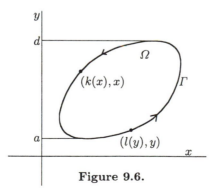

Figure 9.6.

The equality

$$\iint\limits_{\Omega} \left(-\frac{\partial P}{\partial y}\right) dxdy = \int_{\Gamma} P dx$$

is obtained in the same way and this completes the proof.

Green's theorem is true for many other sets Ω and the proof usually proceeds by partitioning Ω into sets $(\Omega_i)_i$ and applying the simple case above to each Ω_i (Figure 9.7). Note that each new curve created in partitioning Ω appears as part of the boundary of *two* Ω_i's and each direction along these new curves appears precisely once. Hence, when we apply (9.2) to each Ω_i and add them together the integrals along the newly created curves cancel and we are left with a line integral over the original curve Γ.

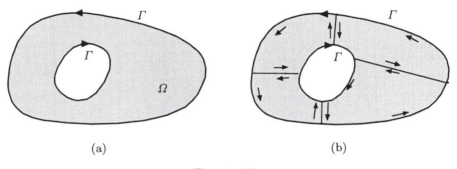

(a)   (b)

Figure 9.7.

A glance at Figure 9.7(a) shows that we no longer have an anticlockwise oriented boundary. In our new situation a *finite* number of piecewise smooth curves form the boundary of Ω and as we proceed along Γ in the given direction the open set Ω lies on the *left-hand side*. With these modifications Green's theorem is still true. We are, of course, always assuming that P and Q are nice smooth functions.

Example 9.3

We wish to calculate

$$I = \int_{\Gamma} (5 - xy - y^2)dx + (2xy - x^2)dy$$

where Γ is the boundary of the unit square $[0,1] \times [0,1]$ in $\mathbb{R}^2$. By Green's theorem

$$
\begin{aligned}
I &= \int_0^1 \int_0^1 \left(\frac{\partial}{\partial x}(2xy - x^2) - \frac{\partial}{\partial y}(5 - xy - y^2)\right)dxdy \\
&= \int_0^1 \int_0^1 (2y - 2x + x + 2y)dxdy \\
&= \int_0^1 \int_0^1 (4y - x)dxdy \\
&= \left(\int_0^1 4ydy\right)\left(\int_0^1 dx\right) - \left(\int_0^1 dy\right)\left(\int_0^1 xdx\right) \\
&= \left.\frac{4y^2}{2}\right|_0^1 \left.x\right|_0^1 - \left.y\right|_0^1 \left.\frac{x^2}{2}\right|_0^1 \\
&= 2 - \frac{1}{2} = \frac{3}{2}.
\end{aligned}
$$

The change of variables rule for double integrals can be considered as a special case of the same rule for triple integrals and this is discussed in Chapter 14.

EXERCISES

9.1 Find $\iint_U x^2 \sin^2 y\, dxdy$ where $A = \{(x,y) \in \mathbb{R}^2, 0 < x < 1, 0 < y < \pi/4\}$.

9.2 Evaluate

(a) $\iint_U x\cos(x + y)dxdy$ where U is the subset of $\mathbb{R}^2$ bounded by the triangle with vertices $(0,0)$, $(\pi,0)$ and (π,π).

(b) $\iint_U (x^2 + y^2)dxdy$ where $U = \{(x,y) \in \mathbb{R}^2; x^2 + y^2 \le 2y\}$.

(c) $\iint_U \frac{y^2}{x^2}dxdy$ where U is the region bounded by $y = x$, $y = 2$ and $xy = 1$.

9.3 Find the area bounded by the curves $x = y^2$ and $x = 4y - y^2$.

9.4 If Γ is a closed anticlockwise directed curve in $\mathbb{R}^2$ with interior Ω and $\mathbf{F}$ is a smooth vector field on an open set containing $\Omega \cup \Gamma$ show that Green's theorem is equivalent to

$$\iint_\Omega \text{div}(\mathbf{F}) = \int_\Gamma (\mathbf{F} \cdot \mathbf{n}) ds$$

where $\mathbf{n}$ is the outward normal to Γ.
If f is harmonic on Ω, i.e.

$$\frac{\partial^2 f}{\partial x^2} + \frac{\partial^2 f}{\partial y^2} = 0$$

and continuous on $\Omega \cup \Gamma$ show that $\int_\Gamma (\nabla f \cdot \mathbf{n}) ds = 0$.

9.5 By reversing the order of integration find

(a) $\int_0^6 \left\{ \int_{x/3}^2 e^{y^2} dy \right\} dx$

(b) $\int_0^4 \left\{ \int_{y/2}^{\sqrt{y}} e^{y/x} dx \right\} dy$

9.6 Find the volume of

$$\{(x, y, z) \in \mathbb{R}^3; x^2 + y^2 \leq 1, x \geq 0, y \geq 0, 0 \leq z \leq 1 - xy\}.$$

9.7 If Ω is an open set enclosed by the anticlockwise directed curve Γ show, using Green's theorem, that

$$\text{Area } (\Omega) = \int_\Gamma x dy = \int_\Gamma -y dx = \frac{1}{2} \int_\Gamma x dy - y dx.$$

Using one of these line integrals find the area of the interior of the ellipse $(x/4)^2 + (y/5)^2 = 1$.

9.8 Verify Green's theorem for the following integrals:

(a) $\int_\Gamma xy^2 dx + 2x^2 y dy$, Γ is the ellipse $4x^2 + 9y^2 = 36$

(b) $\int_\Gamma (x^2 + 2y^3) dy$, Γ is the circle $(x - 2)^2 + y^2 = 4$

(c) $\int_\Gamma 2x^2 y^2 dx - 3yx dy$, Γ is the square bounded by the lines $x = 3$, $x = 5$, $y = 1$, $y = 4$

where Γ is always directed in an anticlockwise direction.

Parametrised Surfaces in $\mathbb{R}^3$

Summary. *We discuss theoretical and practical approaches to parametrizing a surface in* $\mathbb{R}^3$.

In this chapter we begin a systematic study of surfaces in $\mathbb{R}^3$, a topic which will occupy the remaining chapters of this book. We begin with an informal discussion of the background we bring to this investigation and reveal our general intentions.

Daily we encounter most of the classical Euclidean surfaces in $\mathbb{R}^3$ such as a *sphere* (football, globe), *cone* (ice-cream cone), *ellipsoid* (egg, American football), *cylinder* (jar, can), *plane* (floor, wall, ceiling) and the non-Euclidean *torus* (doughnut, tube). Many of our examples involve these surfaces and a solid geometric understanding of these should be deliberately cultivated. Initially these surfaces are merely subsets of $\mathbb{R}^3$ which have a certain recognisable shape. We established mathematical contact with them as level sets and graphs of functions. Our studies were confined to the classical problem in differential calculus of finding the maximum and minimum of a scalar-valued function over a surface and for this we introduced the geometric concepts of *tangent plane* and *normal line*. This essentially summarises the formal knowledge we have acquired but we also have at our disposal a range of mathematical ideas and techniques which, if not directly applicable, will often hint at the way forward.

Our plan is to develop integration theory on surfaces and to investigate the geometry of surfaces. Keep in mind the following topics as particularly relevant to the theories we develop:

(a) integration over open subsets of $\mathbb{R}^2$ (Chapter 9)

(b) integration along directed curves in $\mathbb{R}^2$ and $\mathbb{R}^3$ (Chapter 6)

(c) the geometry of curves in $\mathbb{R}^2$ and $\mathbb{R}^3$ (Chapters 7 and 8).

For instance an open subset U in $\mathbb{R}^2$ can be considered, or identified with, a surface $\widetilde{U}$ in $\mathbb{R}^3$ by means of the mapping

$$(x, y) \in U \longrightarrow (x, y, 0) \in \widetilde{U}.$$

Thus any theory of integration over a surface will extend integration theory over open subsets of $\mathbb{R}^2$. We can say more since parametrizations (Definition 10.2) are a means of identifying a simple surface with an open subset of $\mathbb{R}^2$ and this eventually allows us to reduce integration over a surface to integration over an open subset of $\mathbb{R}^2$ in much the same way that we reduced line integrals in $\mathbb{R}^n$ to integration over an interval in $\mathbb{R}$ (the key word in this sentence is unfortunately "eventually"). Thus our theory of integration over surfaces is obtained by combining and developing techniques already used in (a) and (b). We shall see later how (b) and (c) and, in particular, the methods used to derive the Frenet–Serret equations, can be extended to investigate the geometry of surfaces in $\mathbb{R}^3$.

We begin our formal study of surfaces by defining the concept of a para–metrised surface in $\mathbb{R}^3$. Our definition is mathematically simple, a good starting point, and carries us a long way but does have certain inadequacies that we discuss as we proceed.

Definition 10.1

A parametrised surface in $\mathbb{R}^3$ consists of a pair (S, ϕ) where S is a subset of $\mathbb{R}^3$ and ϕ is a bijective mapping from an open subset of $\mathbb{R}^2$ onto S such that the following conditions hold:

(i) ϕ has derivatives of all orders (we say that ϕ is smooth or $\mathcal{C}^\infty$)

(ii) $\phi_x \times \phi_y \neq 0$ at all points.

Condition (ii) is the analogue of $P' \neq 0$ for a parametrised curve in $\mathbb{R}^3$ and is *equivalent* to the requirement that ϕ_x and ϕ_y are linearly independent vectors at all points (see Figure 10.1).

Definition 10.2

A simple surface in $\mathbb{R}^3$ is a subset S of $\mathbb{R}^3$ for which there exists a mapping ϕ such that (S, ϕ) is a parametrised surface. We call ϕ a parametrization of S.

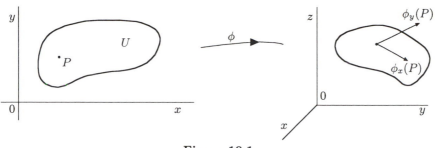

Figure 10.1.

One notable difference between the above definitions and the corresponding definition of directed curve (Definition 5.1) is the change from a *closed* interval in $\mathbb{R}$ to an *open* subset of $\mathbb{R}^2$ for the domain of parametrization. For a directed curve the inclusion of end points in the domain of definition leads to a sense of *direction* along the curve and we will need an analogous concept, a sense of *orientation*, to develop vector-valued integration theory over a surface. However, a sense of direction along a curve can also be obtained by using tangent vectors at interior points of the curve and we develop the concept we require for surfaces by using the interior of the surface.

Unfortunately, many of the classical Euclidean surfaces, e.g. the sphere, are not simple surfaces but, fortunately, for many practical purposes, e.g. the calculation of surface area and the evaluation of surface integrals, they may be considered as simple surfaces. We will make this precise later and also define a general surface in $\mathbb{R}^3$.

We now examine three specific examples – *graphs, surfaces of revolution* and the classical *ellipsoid*. These, although apparently rather limited, appear in many different contexts and we allow them to divert us to essential ideas associated with any parametrised surface.

Example 10.1

(see Example 7.1) Let $f: U \subset \mathbb{R}^2 \to \mathbb{R}$ denote a smooth function defined on the open subset U of $\mathbb{R}^2$ and let S denote the *graph* of f, i.e.

$$S = \{(x, y, f(x, y)) : (x, y) \in U\}.$$

The form of S immediately gives a parametrization ϕ defined by

$$\phi(x, y) = (x, y, f(x, y))$$

with domain U. Since $\phi_x = (1, 0, f_x)$ and $\phi_y = (0, 1, f_y)$ we have

$$\phi_x \times \phi_y = \begin{vmatrix} \mathbf{i} & \mathbf{j} & \mathbf{k} \\ 1 & 0 & f_x \\ 0 & 1 & f_y \end{vmatrix} = (-f_x, -f_y, 1).$$

As the final coordinate of $\phi_x \times \phi_y$ is always 1 we have $\phi_x \times \phi_y \neq 0$. If $\phi(x_1, y_1) = \phi(x_2, y_2)$ then

$$(x_1, y_1, f(x_1, y_1)) = (x_2, y_2, f(x_2, y_2)).$$

Hence $x_1 = x_2$ and $y_1 = y_2$ and so ϕ is injective. Clearly $\phi(U) = S$.

As a particular example consider the unit sphere $S : x^2 + y^2 + z^2 = 1$. We have $z^2 = 1 - x^2 - y^2$ and $z = \pm(1 - x^2 - y^2)^{1/2}$. Hence the *upper hemisphere* of S is the graph of the function

$$g(x, y) = (1 - x^2 - y^2)^{1/2}$$

and the mapping $\widetilde{g} \colon (x, y) \to (x, y, (1 - x^2 - y^2)^{1/2})$ is a parametrization of the upper hemisphere. The domain of $\widetilde{g}$ is the disc $\{(x, y) \in \mathbb{R}^2 : x^2 + y^2 < 1\}$. To visualise this physically imagine the hemisphere as a dome S and take U to be the floor of the dome. If you can now accept yourself as a point p on the floor and look directly upwards your eyes will focus on a unique point q of the dome S (Figure 10.2). Moving around you will be *identifying* the points of U with points of the surface. The floor U is the domain of the parametrization while the parametrization itself is the mapping which takes p to q. The fact that you are always looking in the same direction means that the identification or parametrization is given by a set of *parallel* lines (Figure 10.2) and since parallel lines never meet the mapping is injective or one-to-one. Since each point on the dome is hit by an arrow we also have an onto or surjective mapping and thus a bijective mapping. In general any natural identification of a flat set with a surface by means of parallel lines will lead to a parametrization (see Example 11.2).

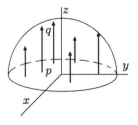

Figure 10.2.

A point P in $\mathbb{R}^2$ can be identified by means of its distance r from the origin and the angle θ between the positive x-axis and the vector OP (Figure 10.3).

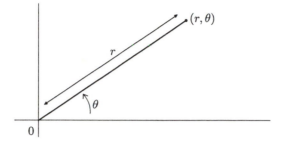

Figure 10.3.

We call (r, θ) the *polar coordinates* of P. Polar coordinates are particularly useful when dealing with circles and discs and with functions involving the expression $x^2 + y^2$ in Cartesian coordinates. Unfortunately, the correspondence

$$(r, \theta) \longrightarrow (x, y) = (r \cos \theta, r \sin \theta)$$

is not bijective but we get around this difficulty by removing a small portion of the domain. The domain of the parametrization of the hemisphere S given above is the disc $x^2 + y^2 < 1$ and suggests the use of polar coordinates. Since

$$(1 - x^2 - y^2)^{1/2} = (1 - r^2 \cos^2 \theta - r^2 \sin^2 \theta)^{1/2} = (1 - r^2)^{1/2}$$

the mapping

$$F \colon (r, \theta) \in (0, 1) \times (-\pi, \pi) \to \left(r \cos \theta, r \sin \theta, (1 - r^2)^{1/2}\right)$$

is a bijective mapping onto $S \setminus \Gamma$ where Γ is the set

$$\{(-r, 0, (1 - r^2)^{1/2}), \, 0 \leq r < 1\}.$$

On $(0, 1) \times (-\pi, \pi)$

$$F_r \times F_\theta = \left(\frac{r^2 \cos \theta}{(1 - r^2)^{1/2}}, \frac{r^2 \sin \theta}{(1 - r^2)^{1/2}}, r\right)$$

is non-zero and we have found a second parametrization of (almost) the whole hemisphere.

Example 10.2

Let $P(t) = \big(x(t), y(t)\big)$, $t \in [a, b]$ denote a directed curve Γ in $\mathbb{R}^2$. We suppose $y(t) > 0$ for all t. The surface obtained by revolving this curve about the x-axis is called the *surface of revolution* of P about the x-axis (Figure 10.4).

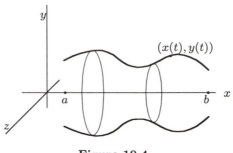

Figure 10.4.

Consider a typical point on the circle obtained by rotating the point $(x(t), y(t))$. The first coordinate remains unchanged. The second coordinate will generate a circle in the (y, z)-plane with centre $(0,0)$ and radius $y(t)$. Using the standard parametrization of the circle we see that a typical point on this circle has coordinates $(y(t) \cos \theta, y(t) \sin \theta)$. Putting these together we obtain a parametrization

$$(t, \theta) \longrightarrow (x(t), y(t) \cos \theta, y(t) \sin \theta)$$

where $t \in (a, b)$ and $\theta \in (0, 2\pi)$. Now

$$\phi_t = (x'(t), y'(t) \cos \theta, y'(t) \sin \theta)$$

and

$$\phi_\theta = (0, -y(t) \sin \theta, y(t) \cos \theta).$$

Hence

$$\phi_t \times \phi_\theta = \begin{vmatrix} \mathbf{i} & \mathbf{j} & \mathbf{k} \\ x'(t) & y'(t) \cos \theta & y'(t) \sin \theta \\ 0 & -y(t) \sin \theta & y(t) \cos \theta \end{vmatrix}$$

$$= (y'(t)y(t), -x'(t)y(t) \cos \theta, -x'(t)y(t) \sin \theta).$$

Since

$$\|\phi_t \times \phi_\theta\| = ((y')^2 y^2 + (x')^2 y^2 \cos^2 \theta + (x')^2 y^2 \sin^2 \theta)^{1/2}$$

$$= y(t)(y'(t)^2 + x'(t)^2)^{1/2} = y(t)\|P'(t)\| \neq 0$$

we have $\phi_t \times \phi_\theta \neq 0$. Since the mapping ϕ is bijective as long as we do not include θ and $\theta + 2\pi$ in the domain we let $U = (a, b) \times (0, 2\pi)$ and obtain a bijective mapping onto the surface of revolution with one curve, the original or *profile* curve, removed. Many classical surfaces may be realised as surfaces of revolution; e.g. the cone (Figure 10.5(a)), cylinder (Figure 10.5(b)) and sphere (Figure 10.5(c)).

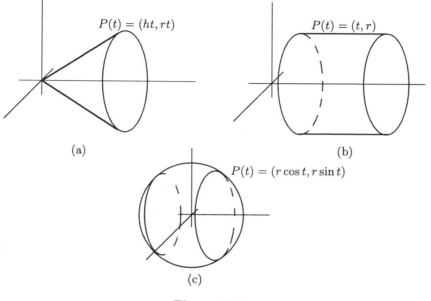

(a)

(b)

$$P(t) = (ht, rt) \qquad P(t) = (t, r)$$

$$P(t) = (r\cos t, r\sin t)$$

(c)

Figure 10.5.

These induce the following parametrizations:

$$
\begin{array}{ll}
\textit{Cone} & (t, \theta) \longrightarrow (ht, rt\cos\theta, rt\sin\theta), \\
\textit{Cylinder} & (t, \theta) \longrightarrow (t, r\cos\theta, r\sin\theta), \\
\textit{Sphere} & (t, \theta) \longrightarrow (r\cos t, r\sin t\cos\theta, r\sin t\sin\theta).
\end{array}
$$

We parametrise, in Example 11.1 the *torus* as a surface of revolution.

We now examine in considerable detail a particular surface – the standard ellipsoid. This includes the sphere as a special case. We use this example as an excuse to explore and comment on many of the practical and theoretical considerations that arise in studying any surface and to shed some light on the necessity and significance of later developments. In studying this example the reader should keep in mind that parametrizations are nothing more than coordinate systems, that we are always trying to interpret mathematical facts geometrically and constantly attempting to articulate mathematically our geometric observations.

Example 10.3

We consider the ellipsoid

$$\frac{x^2}{a^2} + \frac{y^2}{b^2} + \frac{z^2}{c^2} = 1 \,.$$

This specialises to the sphere of radius a centred at the origin when $a = b = c$. Two of the most frequently used coordinate systems on the sphere are the *spherical polar coordinates* and *geographical coordinates*. These, as we shall see, are closely related. Spherical polar coordinates are more popular in the mathematical literature but since geographical coordinates are in everyone's normal experience we devote more time to them here.

The representation of the ellipsoid as a level set of a sum of squares suggests we use the elementary identity $\sin^2 \theta + \cos^2 \theta = 1$ to develop our parametrization. Rewriting the formula for the ellipsoid as a sum of *two* squares we get

$$\left(\sqrt{\frac{x^2}{a^2} + \frac{y^2}{b^2}} \right)^2 + \left(\frac{z}{c} \right)^2 = 1 \,.$$

Now let

$$\sqrt{\frac{x^2}{a^2} + \frac{y^2}{b^2}} = \cos \theta \quad \text{and} \quad \frac{z}{c} = \sin \theta \,.$$

Hence $z = c \sin \theta$ and

$$\frac{x^2}{a^2} + \frac{y^2}{b^2} = \cos^2 \theta, \quad \text{i.e.} \quad \left(\frac{x}{a \cos \theta} \right)^2 + \left(\frac{y}{b \cos \theta} \right)^2 = 1 \,.$$

A further similar substitution gives

$$\frac{x}{a \cos \theta} = \cos \psi \quad \text{and} \quad \frac{y}{b \cos \theta} = \sin \psi \,.$$

This implies $x = a \cos \theta \cos \psi$ and $y = b \cos \theta \sin \psi$. Our parametrization, denoted by F, has the form

$$F : U \longrightarrow S \text{ (ellipsoid)}$$
$$(\theta, \psi) \longrightarrow (a \cos \theta \cos \psi, b \cos \theta \sin \psi, c \sin \theta)$$

but we must specify U, check that F is bijective and see that conditions (i) and (ii) are satisfied. Geometrically we have the following diagram (Figure 10.6) in the case of the sphere $a = b = c$.

Thus we project the point $P = (x, y, z)$ onto the xy-plane to get the point $Q = (x, y, 0)$. The angle θ is the angle between the vectors OQ and OP and hence $\theta = \tan^{-1}(z \,/\, (x^2 + y^2)^{1/2})$. The level set of the surface through P parallel to the xy-plane is a circle (remember we are just considering a sphere

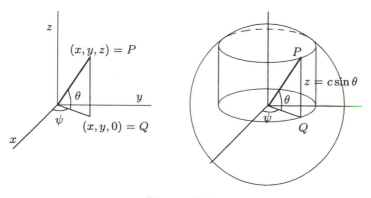

Figure 10.6.

here) and this is also projected onto the xy-plane and the angle ψ is obtained by using *polar coordinates* on this circle. Hence $\psi = \tan^{-1}(y/x)$. From Figure 10.6 it is clear that θ ranges over the interval $(-\pi/2, \pi/2)$ and ψ over the interval $(0, 2\pi)$. Clearly the smoothness condition (i) is satisfied no matter what domain U we choose for F. In many examples condition (ii) is easily checked and formal identification of the range of F is obtained using a diagram. In this example, however, we adopt a more analytic approach in verifying that F is a parametrization.

We have
$$F_\theta = (-a \sin\theta \cos\psi, -b \sin\theta \sin\psi, c \cos\theta)$$
and
$$F_\psi = (-a \cos\theta \sin\psi, b \cos\theta \cos\psi, 0).$$

Hence,

$$F_\theta \times F_\psi = \begin{vmatrix} \mathbf{i} & \mathbf{j} & \mathbf{k} \\ -a \sin\theta \cos\psi & -b \sin\theta \sin\psi & c \cos\theta \\ -a \cos\theta \sin\psi & b \cos\theta \cos\psi & 0 \end{vmatrix}$$

$$= (-bc \cos^2\theta \cos\psi, -ac \cos^2\theta \sin\psi, -ab \cos\theta \sin\theta)$$

$$= -abc \cos\theta \left(\frac{\cos\theta \cos\psi}{a}, \frac{\cos\theta \sin\psi}{b}, \frac{\sin\theta}{c} \right). \tag{10.1}$$

If $a = b = c$ then

$$\|F_\theta \times F_\psi\| = a^2 \cos\theta (\cos^2\theta \cos^2\psi + \cos^2\theta \sin^2\psi + \sin^2\theta)^{1/2}$$
$$= a^2 \cos\theta.$$

Before proceeding we make a brief observation which we develop in more detail in Chapters 16–18. If we fix ψ and let θ vary we obtain a mapping

$$\theta \longrightarrow F(\theta, \psi)$$

where θ ranges over an interval in $\mathbb{R}$. This defines a directed curve which lies in S with tangent F_θ. Similarly F_ψ is tangent to the curve $\psi \to F(\theta, \psi)$ in S. These curves are called *coordinate curves* (of the parametrization). The vectors F_θ and F_ψ lie in the tangent space of S at P and are called *tangent vectors*. Condition (ii) says that they span the tangent space and the following relationship holds in *any* surface S.

$$\{\text{The tangent space of } S \text{ at } P\} = \left\{ \begin{array}{c} \text{set of all tangents at } P \text{ of all curves} \\ \text{in } S \text{ which pass through } P \end{array} \right\}$$

$$= \left\{ \begin{array}{c} \text{all linear combinations spanned} \\ \text{by the partial derivatives at } P \\ \text{of any parametrization.} \end{array} \right\}$$

Moreover, since $F_\theta \times F_\psi \neq 0$, the vector $F_\theta \times F_\psi$ is perpendicular to the tangent space and hence parallel to the normal. In our case the surface S is the level set $g^{-1}(0)$ where

$$g(x, y, z) = \frac{x^2}{a^2} + \frac{y^2}{b^2} + \frac{z^2}{c^2} - 1 \,.$$

We have seen in Chapter 2 that the normal is parallel to ∇g and by inspecting (10.1) we see that $\nabla g(P)$ is parallel to $F_\theta(P) \times F_\psi(P)$. This observation can be used to partially check calculations and, in some cases, may even be used to avoid calculating $F_\phi \times F_\psi$. We have thus related our earlier concepts of *tangent plane* and *normal* with terms which can be calculated from any parametrization.

We turn now to find a domain U for F and to verify that all requirements necessary for a parametrization are satisfied by (F, U). Initially, it is usually better to choose the domain of parametrization to be as large and as simple as possible. These twin aims may not always be compatible with the requirements for a parametrization or, indeed, with one another so some judgement is necessary and this usually comes with experience.

Since we require $F_\theta \times F_\psi \neq 0$ we see from (10.1) that we must have $\cos\theta \neq 0$. A natural choice is the interval $-\pi/2 < \theta < \pi/2$ and this is also suggested by Figure 10.6. Since $\sin^2\psi + \cos^2\psi = 1$ we cannot have $\cos\psi$ and $\sin\psi$ equal to zero for the same value of ψ and thus the condition $\cos\theta \neq 0$ alone implies $F_\theta \times F_\psi \neq 0$. Since both the sine and cosine functions have period 2π the length of the interval of definition of ψ *cannot* be greater than 2π. The interval $(0, 2\pi)$ is a natural choice and leads to

$$U = \{(\theta, \psi) : -\pi/2 < \theta < \pi/2, \ 0 < \psi < 2\pi\}$$

as our domain for F. Clearly the set U is open. If $F(\theta_1, \psi_1) = F(\theta_2, \psi_2)$ then, comparing final coordinates, we get $\sin\theta_1 = \sin\theta_2$. Since the sine function is injective on $(-\pi/2, \pi/2)$ this implies $\theta_1 = \theta_2$. Moreover, $\cos\theta \neq 0$ on $(-\pi/2, \pi/2)$

and comparing the first two coordinates of F we see that $\cos\psi_1 = \cos\psi_2$ and $\sin\psi_1 = \sin\psi_2$. By considering the graphs of both functions (Figure 10.7) we see immediately that F is injective on U.

We have thus shown that F is a parametrization of its range or image, $F(U)$. By our construction the image of F lies in the ellipsoid. Does it cover the full ellipsoid? If $F(\theta,\psi) = (0,0,c)$ then $\sin\theta = 1$. However, $-\pi/2 < \theta < \pi/2$, and so $(0,0,c) \notin F(U)$ and F does *not* cover the ellipsoid. To find the range of F we proceed as follows. The set U is an *open* rectangle in $\mathbb{R}^2$ and its boundary is the perimeter of this rectangle. The function F has a nice smooth extension from U to its boundary (Figure 10.8) and the image of this boundary is the boundary of the image of F in the surface we are examining.

Since $F(\theta,0) = F(\theta,2\pi) = (-a\cos\theta, 0, c\sin\theta)$, $F(\pi/2,\psi) = (0,0,c)$, $F(-\pi/2,\psi) = (0,0,-c)$, and $-\pi/2 < \theta < \pi/2$ the boundary of U is mapped onto one half of the ellipse $(x/a)^2 + (z/c)^2 = 1$, $y = 0$. Hence F is a parametrization of the ellipsoid with one curve removed, the half ellipse PQR in Figure 10.9. As we develop geometric insight we will find that a good sketch will often lead to a rapid identification of the domain of definition of the parametrization and it will not be necessary to go through the protracted investigation we have just completed.

A different choice of domain would have led to a parametrization which almost certainly would have covered a different part of the ellipsoid and perhaps covered the curve that we missed. However, *no parametrization* will cover the full ellipsoid. This, although intuitively clear, is a highly non-trivial mathematical result.

A simple surface may have *different parametrizations* and depending on the circumstances one may be more useful than another. In Examples 10.1 and 10.2 we derived other parametrizations for the "sphere". An important technique that we discuss in Chapter 14 is how to go from one parametrization to another. Since parametrizations are coordinate systems involving variables this process

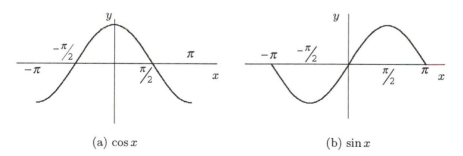

(a) $\cos x$ (b) $\sin x$

Figure 10.7.

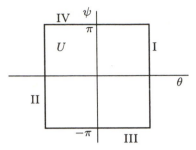

Figure 10.8.

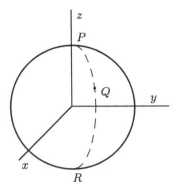

Figure 10.9.

is known as a *change of variable*.

In the above discussion we have seen that ellipsoids and spheres are not simple surfaces. We will see, however, that for the practical purposes we have in mind they may be regarded as simple surfaces. Nevertheless, it is desirable to have a definition of surface in $\mathbb{R}^3$ which implies, for instance, that a sphere is a surface. Such a definition, as we have just seen, requires more than one parametrization.

Definition 10.3

A subset of $\mathbb{R}^3$ is a surface if it can be covered by a collection of (generally overlapping) simple surfaces.

By the *implicit function theorem* a set S of the form $\{X \in U : f(X) = c,\ \nabla f(X) \neq 0\}$, where $f : U(\text{open}) \subset \mathbb{R}^3 \to \mathbb{R}$ is smooth, is a surface and, hence, with this definition an ellipsoid is a surface. We have seen in Chapter 2 that level sets of this type are locally the graphs of smooth functions and hence the method of Example 10.1 can be used, at least in theory, to obtain "local" parametrizations. In this book *all* our results and examples, even when presented for surfaces, follow from studying simple surfaces or objects, like the ellipsoid, which are essentially simple surfaces and we will not call on Definition 10.3 for any technical assistance. This is not as restrictive as it appears and we can include in our study any local geometric property of a surface (see Chapters 16–18). A good understanding of simple surfaces and their limitations is an ideal preparation for the study of more advanced topics such as differentiable manifolds.

We return now to give a geometrical interpretation of the parametrization of the unit sphere, i.e. we consider the ellipsoid with $a = b = c = 1$. A parametrization identifies, with F as the mode of identification, a flat set of points, the open subset U of $\mathbb{R}^2$, with a subset of the surface, $F(U)$, in a one-to-one fashion. So essentially we are taking a sheet of paper U and using F to wrap it around a sphere. The first stage of wrapping turns the sheet of paper into a cylinder with the sphere inside it (Figure 10.10). This identifies the boundary lines I and II with one another (Figure 10.8) and is equivalent mathematically to $F(\theta, 0) = F(\theta, 2\pi)$ for all θ.

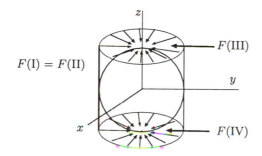

Figure 10.10.

The final steps in the wrapping collapses III onto $(0, 0, c)$ and IV onto $(0, 0, -c)$. Mathematically this says $F(\pi/2, \psi) = (0, 0, c)$ and $F(-\pi/2, \psi) = (0, 0, -c)$ for all ψ. We thus see geometrically that if we *include* the boundary of U in the domain of F then the curve, which we previously missed, is covered twice and F is not one-to-one. If we do *not include* the boundary then this curve is not in the image of F.

Next we discuss parametrizations as *coordinate systems*. A useful initial approach to any such system is to sketch and examine the coordinate curves on the surface. In our case, since we are still looking at the unit sphere, we consider the curves obtained by fixing one of the variables of the function

$$F(\theta, \psi) = (\cos\theta \cos\psi, \cos\theta \sin\psi, \sin\theta), \quad -\pi/2 < \theta < \pi/2, \; 0 < \psi < 2\pi.$$

Since F is bijective, a point on the surface corresponds to a unique pair (θ_1, ψ_1) and $F(\theta_1, \psi_1)$ is the point of intersection of the *coordinate curves* $\theta \rightarrow F(\theta, \psi_1)$, $\psi \rightarrow F(\theta_1, \psi)$. For this reason, the pair (θ, ψ) is often referred to as the *curvilinear coordinates* of the point $F(\theta, \psi)$. Fixing θ is clearly equivalent to taking a fixed value of z and amounts to taking a cross-section of the unit sphere parallel to the xy-plane. Geometrically we get a circle (Figure 10.11).

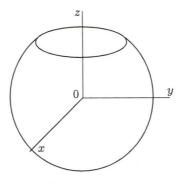

Figure 10.11.

This may also be seen analytically since the mapping $F(\theta, \psi)$ can be written as

$$(0, 0, \sin\theta) + \cos\theta(\cos\psi, \sin\psi, 0).$$

As ψ varies over $(0, 2\pi)$ we get a circle (with one point removed) in a plane parallel to the xy-plane, with centre $(0, 0, \sin\theta)$ on the z-axis, and radius $\cos\theta$. A number of level sets of θ are given in Figure 10.12.

Now fix ψ and let θ range over the interval $-\pi/2 < \theta < \pi/2$. We have

$$F(\theta, \psi) = \cos\theta(\cos\psi, \sin\psi, 0) + \sin\theta(0, 0, 1).$$

Since $(\cos\psi, \sin\psi, 0)$ and $(0, 0, 1)$ are perpendicular unit vectors we get, by Pythagoras' theorem, a semicircle of radius 1. We do not get the full circle because of the restricted range of θ. This is easily seen, either geometrically or from the above formula, to be a semicircle running from $(-1, 0, 0)$ to $(+1, 0, 0)$, i.e. from the bottom to the top of the sphere. In Figure 10.13 we sketch a number of these level sets of ψ.

One can find a ready-made example of this by examining a *globe* representing the Earth. The level sets of θ are the *lines* or *parallels of latitude* with the

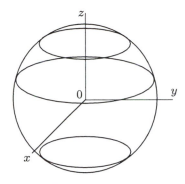

Figure 10.12.

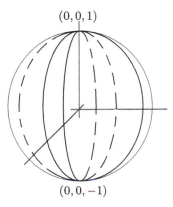

Figure 10.13.

equator corresponding to $\theta = 0$. Instead of $+$ and $-$ the terms North and South are used. The level sets of ψ are called *lines of longitude* or *meridians* with the level set $\psi = 0$ going through Greenwich, England. In place of $+$ and $-$ we use East and West (of Greenwich). The range here is from 0 to 180° while for θ we use 0 to 90°. If we represent the Earth on a map we are unwrapping the sphere, or mathematically, taking the inverse of a parametrization. The most common map (Figure 10.14) is obtained by using *Mercator's projection*. This first unwraps the sphere onto a cylinder in such a way that when the cylinder is unrolled shapes are preserved (but area is distorted – see Exercise 11.8). What happens if we replace the cylinder by a cone?

Note that East meets West on the International Date Line in the middle of the Pacific Ocean. There is a certain ambiguity about this line – is it 180° East or 180° West of Greenwich? This, once more, reflects the fact that we do not obtain a bijective mapping if we include the boundary in the domain of

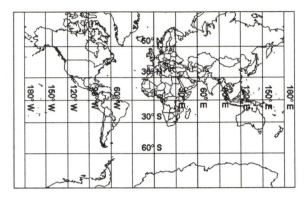

Figure 10.14.

definition of the parametrization.

Different maps result from using different parametrizations or coordinate systems. The mapping F that we have been discussing is called the *geographical coordinate system*. It is as useful and as intuitive a system as the more popular spherical polar coordinate system that we now discuss.

A small change in our method of introducing geographical coordinates leads to the definition of spherical polar coordinates. Recall that, for the ellipsoid, we have

$$\left(\left(\frac{x^2}{a^2} + \frac{y^2}{b^2}\right)^{1/2}\right)^2 + \left(\frac{z}{c}\right)^2 = 1\,.$$

Now we interchange the role of sine and cosine and let

$$\frac{z}{c} = \cos\theta \quad \text{and} \quad \left(\frac{x^2}{a^2} + \frac{y^2}{b^2}\right)^{1/2} = \sin\theta\,.$$

This implies

$$\left(\frac{x}{a\sin\theta}\right)^2 + \left(\frac{y}{b\sin\theta}\right)^2 = 1\,.$$

Let $x/a\sin\theta = \cos\psi$ and $y/b\sin\theta = \sin\psi$. This defines the mapping

$$G(\theta,\psi) = (a\sin\theta\cos\psi, b\sin\theta\sin\psi, c\cos\theta)$$

which are the *spherical polar coordinates* of the point (x,y,z).

From Figure 10.15 we see that $0 < \theta < \pi$ and $0 < \psi < 2\pi$. The angle θ is called the *colatitude* and ψ is sometimes called the *azimuth*. Note that we obtain the same two sets of coordinate curves for geographical and spherical polar coordinates – the labelling is, however, different.

For a sphere of radius r with centre at the origin this gives

$$G(\theta,\psi) = (r\sin\theta\cos\psi, r\sin\theta\sin\psi, r\cos\theta)\,.$$

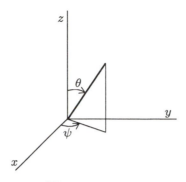

Figure 10.15.

Note that

$$G_\theta \times G_\psi = r^2 \sin\theta(\sin\theta\cos\psi, \sin\theta\sin\psi, \cos\theta)$$

and

$$\|G_\theta \times G_\psi\| = r^2 \sin\theta.$$

If we wish to parametrise a part of the sphere we can still use G – and hence all calculations involving G and its derivatives – but it is necessary to identify the restricted domain on which we are working. Figure 10.15 is useful for this purpose. Suppose, for instance, we wish to parametrise that portion of the sphere of radius 5 with centre at the origin which lies between the planes $z = -3$ and $z = 4$. Figure 10.16(a) and (b) gives the minimum and maximum values of θ, θ_1 and θ_2, and there is no restriction on ψ. Hence our parametrization G has domain

$$U = \{(\theta, \psi) : \sin^{-1}\left(\frac{3}{5}\right) < \theta < \pi - \sin^{-1}\left(\frac{4}{5}\right), \ 0 < \psi < 2\pi\}.$$

From Figure 10.15 we get the spherical polar coordinates in terms of the Cartesian coordinates (x, y, z):

$$r = (x^2 + y^2 + z^2)^{1/2}, \quad \theta = \tan^{-1}\left(\frac{(x^2 + y^2)^{1/2}}{z}\right) \quad \text{and} \quad \psi = \tan^{-1}\left(\frac{y}{x}\right).$$

The parametrization of the sphere can also be used to parametrise the *inverted vertical cone* of height a and base radius a (Figure 10.17(a)).

Note that the angle between the z-axis and the curved surface of the cone is $\pi/4$. If we take the cross-section of the cone of height r, $0 < r < a$, then we obtain a circle of colatitude $\pi/4$ on the sphere of radius $r\sqrt{2}$ (Figure 10.17(b)). The spherical polar coordinates of this circle are

$$\left(r\sqrt{2}\sin\tfrac{\pi}{4}\cos\psi, \ r\sqrt{2}\sin\tfrac{\pi}{4}\sin\psi, \ r\sqrt{2}\cos\tfrac{\pi}{4}\right), \quad 0 < \psi < 2\pi,$$
$$= (r\cos\psi, r\sin\psi, r).$$

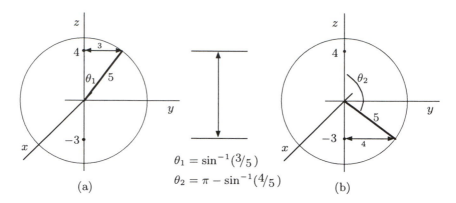

$$\theta_1 = \sin^{-1}(3/5)$$
$$\theta_2 = \pi - \sin^{-1}(4/5)$$

Figure 10.16.

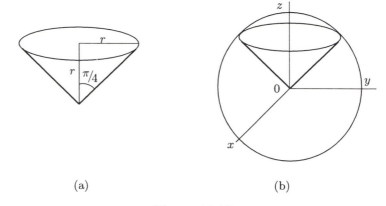

Figure 10.17.

and we have the parametrization

$$(r, \psi) \longrightarrow (r\cos\psi, r\sin\psi, r), \quad 0 < \psi < 2\pi, \ 0 < r < a.$$

This completes our analysis of the ellipsoid and sphere. We remark that many of our comments apply to a wide variety of surfaces.

As a final observation on the material in this chapter we note that many classical surfaces in $\mathbb{R}^3$ may be written as a union of a simple surface and a parametrised curve.

EXERCISES

10.1 By using $\cosh^2 - \sinh^2 = 1$ and $\cos^2 + \sin^2 = 1$ parametrise the surfaces

(a) $\dfrac{x^2}{a^2} + \dfrac{y^2}{b^2} - \dfrac{z^2}{c^2} = 1$

(b) $\dfrac{x^2}{a^2} - \dfrac{y^2}{b^2} - \dfrac{z^2}{c^2} = 1.$

10.2 Show that the following are parametrizations of simple surfaces

(a) $P_1(r, \theta) = (r\cos\theta, r\sin\theta, \theta)$, $\quad 0 < \theta < 2\pi,\ 0 < r < 1$

(b) $P_2(x, \theta) = (x\cos\theta, y\sin\theta, x+y)$, $\quad 0 < x < 1,\ 0 < \theta < 2\pi,\ y \neq 0$

(c) $P_3(u, v) = (u + v, u - v, uv)$, $\quad 0 < u < 1,\ 0 < v < 1$.

10.3 By using a parametrization of an ellipsoid parametrise an inverted cone of height h and base radius r.

10.4 Use geographical and spherical polar coordinates to parametrise that portion of the sphere $x^2 + y^2 + z^2 = r^2$ which lies in the first octant $\{(x, y, z) : x \geq 0, y \geq 0, z \geq 0\}$.

10.5 Let $S = \{(x, y, z) : x^2 + y^2 + z^2 = 1\}$ denote the unit sphere in $\mathbb{R}^3$. If $(u, v) \in \mathbb{R}^2$ then the line determined by $(u, v, 0)$ and $(0, 0, 1)$ intersects S in a point other than $(0, 0, 1)$. Let $\phi(u, v)$ denote this point. Find the coordinates of $\phi(u, v)$ in terms of u and v. Show that $\phi(\mathbb{R}^2)$ is a simple surface parametrised by ϕ.

10.6 Find the length of the portion of the curve $u = v$ lying in the surface parametrised by

$$\phi(u, v) = (u\cos v, u\sin v, u\sqrt{3})$$

which lies between the planes $z = 0$ and $z = 2\sqrt{3}$.

10.7 Let $f(x, y, z) = x + xy + yz$ and let $S = f^{-1}(1)$. Find two parametrised surfaces which lie in S and cover S.

10.8 Consider a line L initially lying along the positive y-axis and attached orthogonally to the z-axis. If L moves along the z-axis with constant speed a and at the same time rotates about the z-axis with constant speed b show that it sweeps out a surface parametrised by

$$\phi(u, t) = (u\sin bt, u\cos bt, at), \quad u \geq 0,\ t \geq 0.$$

11
Surface Area

Summary. *We define and calculate surface area.*

We follow the method used in Chapter 5 to calculate the length of a curve in order to define the (surface) area of a simple surface S in $\mathbb{R}^3$. Let $\phi: U \to S$ denote a parametrization of S where U is an open subset of $\mathbb{R}^2$. We take a rectangular partition of U (Figure 11.1), find the approximate area of the image of each rectangle in the partition, form a Riemann sum and obtain the surface area as the limit of the Riemann sums.

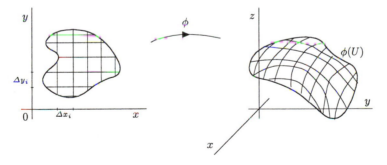

Figure 11.1.

If R denotes a typical rectangle in U (Figure 11.2) then

$$\phi(x + \Delta x, y) - \phi(x, y) \approx \phi_x(x, y)\Delta x$$

123

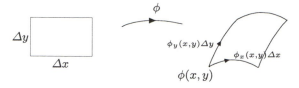

Figure 11.2.

and

$$\phi(x, y + \Delta y) - \phi(x, y) \approx \phi_y(x, y)\Delta y.$$

If $\theta = \theta(x, y)$ is the *angle* between $\phi_x(x, y)$ and $\phi_y(x, y)$ then using the well-known formula for the *area* of a triangle, $\frac{1}{2}ab\sin C$, we get

$$
\begin{aligned}
\text{Area}(\phi(R)) &\approx 2 \cdot \frac{1}{2}\|\phi_x(x, y)\| \cdot \|\phi_y(x, y)\| \cdot |\sin \theta(x, y)|\Delta x \Delta y \\
&= \|\phi_x \times \phi_y(x, y)\|\Delta x \Delta y.
\end{aligned}
$$

If ϕ is integrable over U, and this will be the case if, for instance U is bounded, with smooth boundary, and ϕ has a continuous extension to $\overline{U}$, then

$$
\sum_{i,j} \text{Area}(\phi(R_{ij})) \quad \cong \quad \sum_{i,j} \|\phi_x \times \phi_y(x_i, y_j)\|\Delta x_i \Delta y_j
$$

$$
\longrightarrow \iint_U \|\phi_x \times \phi_y\|dxdy
$$

as we take finer and finer partitions of U. Hence the *surface area* of S, $A(S)$, has the form

$$A(S) = \iint_U \|\phi_x \times \phi_y\|dxdy$$

and is calculated using a parametrization. In general, a surface will admit many different parametrizations but, as we will see later, they all give the same value for surface area. In this chapter we are using the usual *physical* idea of area and angle. These *non-negative absolute quantities* do not require a sense of direction or orientation on the surface and lead, as we have just seen, to a relatively straightforward form of integration. In the next chapter we require more sophisticated concepts of area and angle to integrate *vector fields* over a surface.

We now obtain another formula for surface area which avoids the cross product. We maintain the notation $\phi \colon U \to S$ for our parametrization and introduce, in their traditional form, three quantities that make regular and important appearances in the remaining chapters of this book. Let

$$
\begin{aligned}
E \text{ (or } E(x,y)) &= \phi_x \cdot \phi_x = \|\phi_x\|^2 \\
F \text{ (or } F(x,y)) &= \phi_x \cdot \phi_y \\
G \text{ (or } G(x,y)) &= \phi_y \cdot \phi_y = \|\phi_y\|^2 .
\end{aligned}
$$

We have

$$
\|\phi_x \times \phi_y\|^2 = \|\phi_x\|^2 \|\phi_y\|^2 \sin^2 \delta
$$

where δ is the angle between ϕ_x and ϕ_y. Hence

$$
\cos \delta = \frac{\phi_x \cdot \phi_y}{\|\phi_x\| \cdot \|\phi_y\|}
$$

and

$$
\begin{aligned}
\|\phi_x \times \phi_y\|^2 &= \|\phi_x\|^2 \|\phi_y\|^2 (1 - \cos^2 \delta) \\
&= \|\phi_x\|^2 \|\phi_y\|^2 \left(1 - \frac{(\phi_x \cdot \phi_y)^2}{\|\phi_x\|^2 \|\phi_y\|^2}\right) \\
&= \|\phi_x\|^2 \|\phi_y\|^2 - (\phi_x \cdot \phi_y)^2 = EG - F^2 .
\end{aligned}
$$

This gives the following useful formula for surface area

$$
A(S) = \iint\limits_{U} \sqrt{EG - F^2}\, dx dy .
$$

Figure 11.2 shows how E, F and G quantify the *distortion* of a rectangle by the parametrization. The stretching or contraction of the sides is measured by E and G while F measures the change in angle. Thus we see that *shape* is preserved if $E = G$ and $F = 0$ while (relative) *area* is preserved if $EG - F^2$ is constant. For many important parametrizations, including geographical and spherical polar coordinates, $F = 0$. This implies that angles between curves are preserved and, in particular, parallels (of latitude) and meridians (of longitude) cross one another at right angles. For geographical coordinates on a sphere of radius r, $E = r^2$ and $G = r^2 \cos^2 \theta$ and hence neither shape nor area are preserved. On the Equator, where $\theta = 0$, we have $E = G = r^2$ but as one moves towards the North and South Poles, $\theta \to \pm\pi/2$ and $G \to 0$ while $E = r^2$. Consequently, near the Equator shape is fairly well preserved but as one moves towards the polar regions it becomes more and more distorted. Mercator's projection is a *modification* of geographical coordinates – notice how the lines of latitude in Figure 10.14 are not equally spaced in order to preserve shape. If the cylinder in Mercator's construction is replaced by a cone we obtain *Lambert's equal area projection* which preserves (relative) area but distorts shape and distance.

In the previous chapter we noted that many well-known surfaces S could be written in the form $S = S_1 \cup \Gamma$ where S_1 is a simple surface and Γ is a curve parametrised by a *smooth* function on a closed interval. We see now that the surface area of Γ is zero and from this conclude that $A(S) = A(S_1)$ and thus we may use a parametrization of S_1 to calculate the surface area of S. To show $A(\Gamma) = 0$ enclose Γ in a finite union of rectangles of small width ε (Figure 11.3).

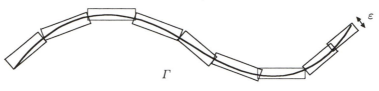

Figure 11.3.

The sum of the lengths of the rectangles is approximately the length of Γ, $l(\Gamma)$. Hence

$$A(\Gamma) \approx \varepsilon \times l(\Gamma) \,.$$

Since ε is arbitrarily small this implies $A(\Gamma) = 0$. This intuitive "proof" can be developed into a rigorous proof using the *mean value theorem* for differentiable functions of one variable. A differentiable parametrization of the curve is *necessary*, although our intuition might suggest otherwise, since there is a famous example of a *square filling curve* parametrised by a continuous function.

Example 11.1

In this example we consider the surface of revolution S parametrised as in Example 10.2. Let $P: [a, b] \to \mathbb{R}^2$ denote a parametrization of the curve to be revolved. Our parametrization ϕ of S is defined on $U = \{(t, \theta) : a < t < b, \, 0 < \theta < 2\pi\}$ and has the form

$$\phi(t, \theta) = \big(x(t), y(t) \cos \theta, y(t) \sin \theta\big)$$

where $P(t) = \big(x(t), y(t)\big)$. We have already shown in Example 10.2 that

$$\|\phi_t \times \phi_\theta\| = y(t)\|P'(t)\|$$

and so the surface area of S is

$$\int_0^{2\pi} \int_a^b y(t)\|P'(t)\|dtd\theta \;=\; \left(\int_0^{2\pi} d\theta\right)\left(\int_a^b y(t)\|P'(t)\|dt\right)$$

$$= 2\pi \int_a^b y(t)\|P'(t)\|dt\,.$$

For the *cone* of height h and (base) radius r, $P(t) = (ht, rt)$, $P'(t) = (h, r)$, $\|P'(t)\| = (h^2 + r^2)^{1/2}$ and the curved surface area is

$$2\pi \int_0^1 rt(r^2 + h^2)^{1/2}dt = 2\pi r(h^2 + r^2)^{1/2} \int_0^1 t\,dt = \pi r(h^2 + r^2)^{1/2}\,.$$

For the *cylinder* of radius r and height h we have $P(t) = (t, r)$, $P'(t) = (1, 0)$ and $\|P'(t)\| = 1$. Hence the curved surface area of the cylinder is

$$2\pi \int_0^h r \cdot 1\,dt = 2\pi r h\,.$$

For the *sphere* of radius r, $P(t) = (r\cos t, r\sin t)$, $P'(t) = (-r\sin t, r\cos t)$ and $\|P'(t)\| = r$ and the surface area of the sphere of radius r is

$$2\pi \int_0^\pi r\sin t \cdot r\,dt = 2\pi r^2(-\cos t)\Big|_0^\pi = 4\pi r^2\,.$$

We now parametrise the *torus* by realising it as a surface of revolution. We adopt a slightly different approach to that of Example 10.2 in order to obtain what is regarded as the standard parametrization. Place a circle of radius r in the (x, z)-plane with centre on the x-axis at a distance b, $r > b$, from the origin and revolve this circle about the z-axis (Figure 11.4).

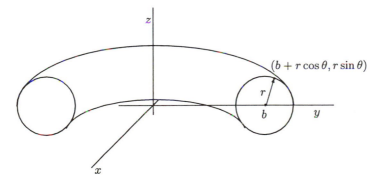

Figure 11.4.

The coordinates of a typical point on the original curve in the (x, z)-plane are $(b + r\cos\theta, r\sin\theta)$. When rotated about the z-axis the third coordinate remains unchanged while the first two coordinates describe a circle of radius $b + r \cos \theta$ about the origin. This gives a parametrization f with formula

$$f(\theta, \psi) = \left((b + r \cos \theta) \cos \psi, (b + r \cos \theta) \sin \psi, r \sin \theta\right)$$

and domain $(0, 2\pi) \times (0, 2\pi)$. Hence

$$
\begin{aligned}
f_\theta &= (-r \sin \theta \cos \psi, -r \sin \theta \sin \psi, r \cos \theta) \\
f_\psi &= \left(-(b + r \cos \theta) \sin \psi, (b + r \cos \theta) \cos \psi, 0\right).
\end{aligned}
$$

By inspection we see that f_θ and f_ψ are perpendicular and hence $F = 0$. We have

$$
\begin{aligned}
E = \| f_\theta \|^2 &= r^2 \sin^2 \theta \cos^2 \psi + r^2 \sin^2 \theta \sin^2 \psi + r^2 \cos^2 \theta \\
&= r^2 \sin^2 \theta (\cos^2 \psi + \sin^2 \psi) + r^2 \cos^2 \theta \\
&= r^2 \sin^2 \theta + r^2 \cos^2 \theta = r^2
\end{aligned}
$$

and

$$
\begin{aligned}
G = \| f_\psi \|^2 &= (b + r \cos \theta)^2 \sin^2 \psi + (b + r \cos \theta)^2 \cos^2 \psi \\
&= (b + r \cos \theta)^2.
\end{aligned}
$$

Hence, $\sqrt{EG - F^2} = r(b + r \cos \theta)$, and

$$
\begin{aligned}
A(\text{Torus}) &= \int_0^{2\pi} \int_0^{2\pi} \sqrt{EG - F^2} \, d\theta \, d\psi \\
&= \int_0^{2\pi} \int_0^{2\pi} r(b + r \cos \theta) \, d\theta \, d\psi \\
&= \left(\int_0^{2\pi} d\psi\right)\left(\int_0^{2\pi} r(b + r \cos \theta) \, d\theta\right) \\
&= 4\pi^2 rb.
\end{aligned}
$$

Some geometric insight might have led you to this answer without *any* calculations or integration (see Example 14.4). The coordinates (θ, ψ) for the torus defined above are called *toroidal polar coordinates*.

Example 11.2

We wish to find the surface area of the portion of the *paraboloid* $z = 4 - x^2 - y^2$ that lies above the xy-plane. This is the graph of the function $f(x, y) = 4 - x^2 - y^2$ and we could find a parametrization using the method in Example 10.1. We prefer, however, to take a more geometric approach.

The standard paraboloid $z = x^2 + y^2$ can be sketched by noting that cross-sections parallel to the xy-plane are circles (Figure 11.5(a)). This surface is turned upside down by taking $z = -x^2 - y^2$ and then moved up 4 units in the direction of the z-axis to give us the original surface $z = 4 - x^2 - y^2$ (Figure 11.5(b)).

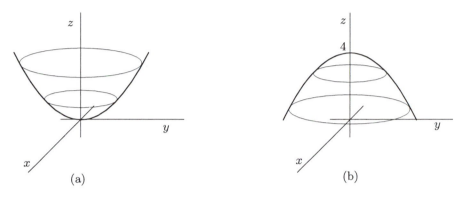

(a)   (b)

Figure 11.5.

The surface cuts the xy-plane when $0 = z = 4 - x^2 - y^2$, i.e. on the circle $x^2 + y^2 = 4$. Clearly the geometry shows that we can project the surface in a one-to-one fashion onto the disc $x^2 + y^2 \leq 4$ and if we reverse this we obtain the parametrization

$$(x, y) \in \{x^2 + y^2 < 4\} \longrightarrow (x, y, 4 - x^2 - y^2) \in \text{ Paraboloid.}$$

We could proceed directly to use this parametrization to compute the surface area. This would require a change of variable in working out the double integral. It is just as easy to initially use a more appropriate parametrization which avoids a later change of variable.

We note that the domain of the above parametrization is a *disc* and the formula for the parametrization function involves $x^2 + y^2$. Either of these on their own suggest polar or spherical coordinates. Since, however, the disc shape and $x^2 + y^2$ only involve the first two coordinates this inclines us towards polar coordinates for the pair (x, y). We use the parametrization

$$f \colon (r, \theta) \longrightarrow (r \cos \theta, r \sin \theta, 4 - r^2), \quad 0 < r < 2 \text{ and } 0 < \theta < 2\pi .$$

We have $f_r = (\cos \theta, \sin \theta, -2r)$, $f_\theta = (-r \sin \theta, r \cos \theta, 0)$. Hence $E = 1 + 4r^2$, $G = r^2$ and since $f_r \cdot f_\theta = 0$ we have $F = 0$. The surface area is

$$\int_0^2 \int_0^{2\pi} \sqrt{EG - F^2} \, dr \, d\theta = \int_0^2 \int_0^{2\pi} r\sqrt{1 + 4r^2} \, dr \, d\theta$$

$$= 2\pi \int_0^2 r\sqrt{1 + 4r^2} \, dr$$

$$= \frac{2\pi}{8} \int u^{1/2} du, \qquad 1 + 4r^2 = u, \ 8r \, dr = du$$

$$= \frac{\pi}{4} \cdot \frac{2}{3} u^{3/2} = \frac{\pi}{6}(1 + 4r^2)^{3/2} \Big|_0^2$$

$$= \frac{\pi}{6}(17^{3/2} - 1).$$

Example 11.3

We wish to find in this example the surface area of that portion of the sphere $x^2 + y^2 + z^2 = 4z$ which lies outside the paraboloid $x^2 + y^2 = 3z$.

For this problem we first get some idea of the geometry and see, once this is reasonably clear, that the parametrization, the domain of the parametrization and even the integration are relatively straightforward.

We have two surfaces, a sphere and a paraboloid. Rewriting the equation for the sphere by completing the square we get

$$x^2 + y^2 + z^2 - 4z + 4 = 4$$

i.e.

$$x^2 + y^2 + (z - 2)^2 = 2^2$$

and this is the sphere of radius 2 with centre $(0, 0, 2)$ (Figure 11.6).

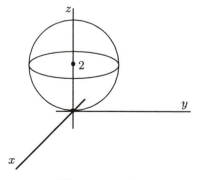

Figure 11.6.

If we take the cross-section of the paraboloid corresponding to the plane $z = c$ we get a circle of radius $\sqrt{3c}$ and so we can sketch the paraboloid

(Figure 11.7). The points of intersection of the two surfaces are identified by letting

$$x^2 + y^2 + z^2 - 4z = 0 = x^2 + y^2 - 3z.$$

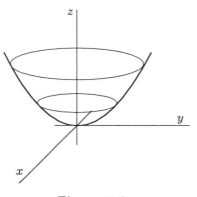

Figure 11.7.

This implies $z^2 - 4z = -3z$, i.e. $z^2 - z = 0$ and $z = 1$ or $z = 0$. When $z = 0$ we get the point $(0,0,0)$ on both surfaces and when $z = 1$ we get the curve $x^2 + y^2 = 3$, i.e. a circle of radius $\sqrt{3}$. This usually indicates where one surface crosses over from the inside to the outside of the other and gives us the working diagram (Figure 11.8).

To find when the sphere is outside the paraboloid it suffices to compare the points (x, y, z_p) and (x, y, z_s) on the paraboloid and sphere, respectively, which project onto the *same* point $(x, y, 0)$ in the (x, y)-plane. From the defining equations of the surfaces we have

$$x^2 + y^2 = 4z_s - z_s^2 = 3z_p.$$

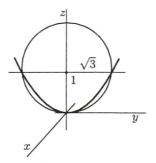

Figure 11.8.

Hence

$$z_s - z_s^2 = 3z_p - 3z_s, \quad \text{i.e.} \quad z_s(1 - z_s) = 3(z_p - z_s).$$

Since $z_s \geq 0$ we have $z_p \geq z_s$ if and only if $1 - z_s \geq 0$, i.e. when $0 \leq z_s \leq 1$. Thus the part of the sphere which lies outside the paraboloid is precisely that portion which lies between the parallel planes $z = 0$ and $z = 1$.

We thus have to find the surface area of the sphere $x^2 + y^2 + (z - 2)^2 = 2^2$ between the planes $z = 0$ and $z = 1$. We use the formula for spherical polar coordinates obtained in Example 10.3 translated so that the origin is at the point $(0, 0, 2)$, i.e. let

$$x = 2\sin\theta\cos\psi, \quad y = 2\sin\theta\sin\psi, \quad z - 2 = 2\cos\theta$$

and it is now only necessary to find the domain of the parametrization. Since $0 \leq z \leq 1$ we consider $-2 \leq z - 2 \leq -1$, i.e. $-2 \leq 2\cos\theta \leq -1$. This implies (see Figure 10.15) $2\pi/3 \leq \theta \leq \pi$ and since there is no restriction on ψ, $0 < \psi < 2\pi$.

Denote this parametrization by f. Then

$$f_\theta = (2\cos\theta\cos\psi, 2\cos\theta\sin\psi, -2\sin\theta)$$

and

$$f_\psi = (-2\sin\theta\sin\psi, 2\sin\theta\cos\psi, 0).$$

Hence

$$\begin{aligned} E &= 4\cos^2\theta\cos^2\psi + 4\cos^2\theta\sin^2\psi + 4\sin^2\theta \\ &= 4(\cos^2\psi + \sin^2\psi)\cos^2\theta + 4\sin^2\theta \\ &= 4\cos^2\theta + 4\sin^2\theta = 4 \end{aligned}$$

and

$$G = 4\sin^2\theta\sin^2\psi + 4\sin^2\theta\cos^2\psi = 4\sin^2\theta.$$

Since $f_\theta \cdot f_\psi = 0$, $F = 0$ and the surface area is

$$\begin{aligned} \int_0^{2\pi}\int_{2\pi/3}^{\pi} 2 \cdot 2\sin\theta \, d\theta d\psi &= \left(4\int_0^{2\pi} d\psi\right)\left(\int_{2\pi/3}^{\pi}\sin\theta d\theta\right) \\ &= 8\pi(-\cos\theta)\Big|_{2\pi/3}^{\pi} = 8\pi(1 - 1/2) = 4\pi. \end{aligned}$$

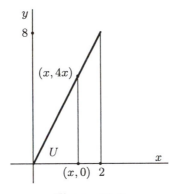

Figure 11.9.

Example 11.4

We calculate the surface area of the level set $x^2 - 2y - 2z = 0$ which projects onto the region U of the xy-plane bounded by the lines $x = 2$, $y = 0$ and $y = 4x$. On the surface S we have $z = \frac{1}{2}x^2 - y$ and the surface is the graph of the function $g(x, y) = \frac{1}{2}x^2 - y$. Using Example 10.1 we obtain the parametrization

$$\phi: (x, y) \in U \longrightarrow (x, y, \frac{1}{2}x^2 - y)$$

and, moreover,

$$\|\phi_x \times \phi_y\| = \|(-g_x, -g_y, 1)\| = \|(x, -1, 1)\| = (x^2 + 2)^{1/2}.$$

Note that only part of the surface lies above the xy-plane. Our surface area is

$$\iint_U (x^2 + 2)^{1/2} dx dy.$$

We integrate first with respect to y and note from Figure 11.9 that y varies from 0 to $4x$ and afterwards integrate with respect to x which varies from 0 to 2. We have

$$
\begin{aligned}
\iint_U (x^2 + 2)^{1/2} dx dy &= \int_0^2 \left\{ \int_0^{4x} (x^2 + 2)^{1/2} dy \right\} dx \\
&= \int_0^2 (x^2 + 2)^{1/2} y \Big|_0^{4x} dx = \int_0^2 4x(x^2 + 2)^{1/2} dx \\
&= 2 \int u^{1/2} du = 2 \frac{u^{3/2}}{3/2} = \frac{4}{3}(x^2 + 2)^{3/2} \Big|_0^2 \\
&= \frac{4}{3}(6^{3/2} - 2^{3/2}) = \frac{8}{3}(3\sqrt{6} - \sqrt{2})
\end{aligned}
$$

where $u = x^2 + 2$, $du = 2x dx$.

EXERCISES

11.1 Calculate the surface area of the paraboloid $z = x^2 + y^2$ which lies between the planes $z = 0$ and $z = 4$.

11.2 Find the area of the surface generated by revolving the curves

(a) $y = x^3$,

(b) $y = x^2$, about the x-axis, where $0 \le x \le 1$.

11.3 Show that the area of the *helicoid* defined by the parametrization

$$P(r, \theta) = (r \cos \theta, r \sin \theta, \theta), \quad 0 < \theta < 2\pi, \ 0 < r < 1$$

is $\pi \left(\sqrt{2} + \log(1 + \sqrt{2}) \right)$.

11.4 If U is open in $\mathbb{R}^2$ and $f : U \to \mathbb{R}$ is a smooth function show

$$\text{Area} \left(\text{graph}(f) \right) = \iint_U \left(1 + \|\nabla f\|^2 \right)^{1/2} dx dy \,.$$

11.5 Find the surface area of the paraboloid in Example 11.3 that lies inside the sphere.

11.6 Find the area of the portion of the surface $z = xy$ which lies inside the cylinder $x^2 + y^2 = a^2$.

11.7 Use the parametrization $(r, \theta) \to (r^2/16, r \cos \theta, r \sin \theta)$ to find the area of the part of $y^2 + z^2 = 16x$ which lies in the first octant ($x \ge 0$, $y \ge 0$, $z \ge 0$) between the planes $x = 0$ and $x = 12$ and inside the cylinder $y^2 = 4x$.

11.8 Let $\phi : U \to S$ denote a parametrization of the simple surface S in $\mathbb{R}^3$. Give some intuitive reasons why ϕ preserves

(a) angles if $F = 0$

(b) relative area if $EF - G^2$ is constant

(c) shape if $E = G$ and $F = 0$.

11.9 Use the Cauchy–Schwarz inequality (Example 3.4) to show that $EG - F^2 \ge 0$ where E, F and G are calculated from any parametrization of a surface.

Table 11.1. Useful Parametrizations.

---

1. Sphere $x^2 + y^2 + z^2 = r^2$;   **geographical coordinates**

$$f(\theta, \psi) = (r \cos \theta \cos \psi, r \cos \theta \sin \psi, r \sin \theta), \ -\pi/2 < \theta < \pi/2, \ 0 < \psi < 2\pi$$

$$f_\theta = (-r \sin \theta \cos \psi, -r \sin \theta \sin \psi, r \cos \theta), \quad f_\psi = (-r \cos \theta \sin \psi, r \cos \theta \cos \psi, 0)$$

$$f_\theta \times f_\psi = -r^2 \cos \theta (\cos \theta \cos \psi, \cos \theta \sin \psi, \sin \theta)$$

$$E = r^2, \ F = 0, \ G = r^2 \cos^2 \theta, \ \|f_\theta \times f_\psi\| = \sqrt{EG - F^2} = r^2 \cos \theta$$

$$\mathbf{n} = -(\cos \theta \cos \psi, \cos \theta \sin \psi, \sin \theta)$$

---

2. Sphere $x^2 + y^2 + z^2$;   **spherical polar coordinates**

$$f(\theta, \psi) = (r \sin \theta \cos \psi, r \sin \theta \sin \psi, r \cos \theta), \quad 0 < \theta < \pi, \ 0 < \psi < 2\pi$$

$$f_\theta = (r \cos \theta \cos \psi, r \cos \theta \sin \psi, -r \sin \theta), \quad f_\psi = (-r \sin \theta \sin \psi, r \sin \theta \cos \psi, 0)$$

$$f_\theta \times f_\psi = r^2 \sin \theta (\sin \theta \cos \psi, \sin \theta \sin \psi, \cos \theta)$$

$$E = r^2, \ F = 0, \ G = r^2 \sin^2 \theta, \ \|f_\theta \times f_\psi\| = \sqrt{EG - F^2} = r^2 \sin \theta$$

$$\mathbf{n} = (\sin \theta \cos \psi, \sin \theta \sin \psi, \cos \theta)$$

---

3. Cylinder $x^2 + y^2 = r^2, \ 0 < z < h$;   **cylindrical polar coordinates**

$$f(\theta, z) = (r \cos \theta, r \sin \theta, z), \quad 0 < \theta < 2\pi, \ 0 < z < h$$

$$f_\theta = (-r \sin \theta, r \cos \theta, 0), \quad f_z = (0, 0, 1)$$

$$f_\theta \times f_z = (r \cos \theta, r \sin \theta, 0)$$

$$E = r^2, \ F = 0, \ G = 1, \ \|f_\theta \times f_z\| = \sqrt{EG - F^2} = r$$

$$\mathbf{n} = (\cos \theta, \sin \theta, 0)$$

---

4. Inverted cone $x^2 + y^2 = z^2, \ 0 < z < a$

$$f(r, \theta) = (r \cos \theta, r \sin \theta, r), \quad 0 < r < a, \ 0 < \theta < 2\pi$$

$$f_r = (\cos \theta, \sin \theta, 1), \quad f_\theta = (-r \sin \theta, r \cos \theta, 0)$$

$$f_r \times f_\theta = (-r \cos \theta, -r \sin \theta, r)$$

$$E = 2, \ F = 0, \ G = r^2, \ \|f_r \times f_\theta\| = \sqrt{EG - F^2} = \sqrt{2} r$$

$$\mathbf{n} = -\frac{1}{\sqrt{2}}(\cos \theta, \sin \theta, -1)$$

Table 11.1. (continued).

---

5. Torus – rotate circle of radius r in xz-plane, centre $(b, 0)$, $b > r$, about z-axis; **toroidal polar coordinates**

$$f(\theta, \psi) = \big((b + r\cos\theta)\cos\psi, (b + r\cos\theta)\sin\psi, r\sin\theta\big), \quad 0 < \theta, \psi < 2\pi$$

$$f_\theta = (-r\sin\theta\cos\psi, -r\sin\theta\sin\psi, r\cos\theta),$$

$$f_\psi = \big(-(b + r\cos\theta)\sin\psi, (b + r\cos\theta)\cos\psi, 0\big)$$

$$f_\theta \times f_\psi = -r(b + r\cos\theta)(\cos\theta\cos\psi, \cos\theta\sin\psi, \sin\theta)$$

$$E = r^2, \ F = 0, \ G = (b + r\cos\theta)^2, \|f_\theta \times f_\psi\| = \sqrt{EG - F^2} = r(b + r\cos\theta)$$

$$\mathbf{n} = -(\cos\theta\cos\psi, \cos\theta\sin\psi, \sin\theta)$$

---

6. Graph of $f: U \subset \mathbb{R}^2 \to \mathbb{R}$

$$F(x, y) = (x, y, f(x, y)), \ (x, y) \in U$$

$$F_x = (1, 0, f_x), \ F_y = (0, 1, f_y), \ F_x \times F_y = (-f_x, -f_y, 1)$$

$$E = 1 + f_x^2, \ F = f_x f_y, \ G = 1 + f_y^2, \|F_x \times F_y\| = \sqrt{EG - F^2} = 1 + f_x^2 + f_y^2$$

$$\mathbf{n} = \frac{(-f_x, -f_y, 1)}{\sqrt{1 + f_x^2 + f_y^2}}$$

---

7. Surface of revolution of Γ parametrised by $P(t) = ((x(t), y(t)), \ a \le t \le b$, $y(t) > 0$ for all t, $\|P'(t)\| \ne 0$

$$f(t, \theta) = \big(x(t), y(t)\cos\theta, y(t)\sin\theta\big), \quad a < t < b, \ 0 < \theta < 2\pi$$

$$f_t = \big(x'(t), y'(t)\cos\theta, y'(t)\sin\theta\big), \quad f_\theta = \big(0, -y(t)\sin\theta, y(t)\cos\theta\big)$$

$$f_t \times f_\theta = \big(y'(t)y(t), -x'(t)y(t)\cos\theta, -x'(t)y(t)\sin\theta\big)$$

$$E = \|P'(t)\|, \ F = 0, \ G = y(t)^2, \|f_t \times f_\theta\| = \sqrt{EG - F^2} = y(t)\|P'(t)\|$$

$$\mathbf{n} = \frac{1}{\|P'(t)\|}\big(y'(t), -x'(t)\cos\theta, -x'(t)\sin\theta\big)$$

12
Surface Integrals

Summary. *We define the integral of a vector field over an oriented surface. Geometrical interpretations are discussed.*

Integrals are used to measure quantities such as length, area, expected value, etc., and as with all measurements, there has to be a unit of measurement. Our basic unit of measurement in integration theory over $\mathbb{R}$ is obtained by assigning the value 1 to the rectangle of height 1 over an interval of length 1 measured from *left to right*. From this we are able to define Riemann sums and afterwards the Riemann integral of a continuous function over a closed interval. The inclusion of "left to right" is crucial for without it we would have an ambiguous definition – and to a mathematician an ambiguous definition is no definition. To emphasise this point we call an interval $[a, b]$ directed from left to right a *positive* interval in $\mathbb{R}$.

Now suppose Γ is a curve in $\mathbb{R}^2$ and, for the sake of simplicity, we suppose that Γ is not closed and that we are interested in defining the integral of a function over Γ. We have seen in Chapter 7 that we can define *two* integrals over Γ since Γ can be directed in two different ways and we have to decide,

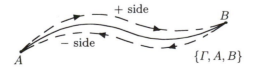

Figure 12.1.

137

before doing any calculations, which interests us. We may think of a curve as having *two sides* – one for each direction of motion – like two escalators side by side, one going up and the other coming down (Figure 12.1). We call one side positive and the other negative. If we are interested in evaluating integrals over $\{\Gamma, A, B\}$ we call Γ directed from A to B the *positive side*.

We use a parametrization to transfer the integral over $\{\Gamma, A, B\}$ to an integral over an interval in $\mathbb{R}$ which we subsequently evaluate. However, the parametrization must be directed correctly. This means that it must map a *positive* interval in $\mathbb{R}$ onto the *positive* side of Γ. Surface integrals are defined in the same way – just step up a dimension.

Consider the unit sphere $x^2 + y^2 + z^2 = 1$ in $\mathbb{R}^3$. It has two *sides* – an inside and an outside. More generally, the surface S which is the level set $\{X \in U : f(X) = 0$ and $\nabla f(X) \neq 0\}$ has $\pm \nabla f(X) / \|\nabla f(X)\|$ as unit normals at each point and we may consider $\nabla f(X) / \|\nabla f(X)\|$ and $-\nabla f(X) / \|\nabla f(X)\|$ as lying on different sides. We *distinguish* between the sides of a surface by *using normals*.

Definition 12.1

An oriented surface in $\mathbb{R}^3$ is a pair $(S, \mathbf{n})$ where S is a surface and $\mathbf{n}$ is a continuous mapping from S into $\mathbb{R}^3$ such that $\mathbf{n}(P)$ is a unit normal to the surface S at P.

We will also use $\mathbf{S}$ to denote an oriented surface. The appearance of $\mathbf{S}$ signals that we have fixed an orientation on the underlying surface S. The notation $\mathbf{S}$ is simpler but if we need to specify the normal or if there is any possibility of confusion we write $(S, \mathbf{n})$.

Clearly if $(S, \mathbf{n})$ is an oriented surface, then $(S, -\mathbf{n})$ is also an oriented surface and the continuity requirement in the choice of normal implies that a *connected surface* can have at most *two orientations*. There exist, however, surfaces which do not admit *any* orientation and integration theory *cannot* be defined over such surfaces. Fortunately, every simple surface admits an orientation. If $\phi\colon U \subset \mathbb{R}^2 \to S$ is a parametrization of S then $\left(S, \phi_x \times \phi_y \,/\, \|\phi_x \times \phi_y\|\right)$ and $\left(S, -\phi_x \times \phi_y \,/\, \|\phi_x \times \phi_y\|\right)$ are two oriented surfaces associated with S and if S is connected these are the only two oriented surfaces associated with S. Given an oriented surface $(S, \mathbf{n})$ we call the side of S containing $\mathbf{n}$ the *positive side* and the side which contains $-\mathbf{n}$ the negative side. In the case of a simple oriented surface $(S, \mathbf{n})$ a parametrization ϕ is said to be *consistent* with the orientation if

$$\frac{\phi_x \times \phi_y}{\|\phi_x \times \phi_y\|} = \mathbf{n}.$$

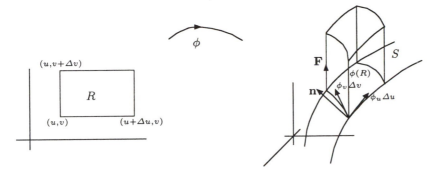

Figure 12.2.

If $\mathbf{F}$ is a continuous vector field on a simple oriented surface $\mathbf{S}$ we define

$$\iint_{\mathbf{S}} \mathbf{F}$$

by using a parametrization (ϕ, U) consistent with the orientation and a process involving Riemann sums similar to that used to define surface area in the previous chapter (see Figure 11.2 and 11.3 for reference). We begin by partitioning the domain U into small rectangles and consider a typical rectangle R in this partition. Now, however, instead of taking the *absolute area* of $\phi(R)$ we take the *(signed) volume* of the *parallelepiped* with base $\phi(R)$ and height determined by $\mathbf{F}$ (see Figure 12.2). We find it convenient to use (u,v) for the variables in the domain of ϕ and (x,y,z) for the range, i.e. $x = x(u,v)$, $y = y(u,v)$ and $z = z(u,v)$.

We use $\mathbf{n}$ as our *positive unit of measurement* perpendicular to the surface. This is reasonable since we are just considering a small portion of the surface $\phi(R)$, which lies approximately in the tangent plane, and $\mathbf{n}(\phi(u,v))$ is perpendicular to the tangent plane at $\phi(u,v)$. The height of the parallelepiped is $\|\mathbf{F}\| \cos\psi = \mathbf{F} \cdot \mathbf{n}$ where ψ is the angle between $\mathbf{F}$ and $\mathbf{n}$ at $\phi(u,v)$ (see Figure 12.3).

We have already seen, in calculating surface area, that the area of $\phi(R)$, the base of the parallelepiped, is $\|\phi_x \times \phi_y\| \Delta x \Delta y$ and hence the volume is

$$(\mathbf{F} \cdot \mathbf{n})\|\phi_u \times \phi_v\| \Delta u \Delta v = (\mathbf{F} \cdot \phi_u \times \phi_v) \Delta u \Delta v \,.$$

Now taking the limit of Riemann sums in the usual way we define

$$\iint_{\mathbf{S}} \mathbf{F} \;=\; \iint_{U} (\mathbf{F} \cdot \mathbf{n})\|\phi_u \times \phi_v\| \, du dv$$

$$\;=\; \iint_{U} (\mathbf{F} \cdot \phi_u \times \phi_v) \, du dv \,.$$

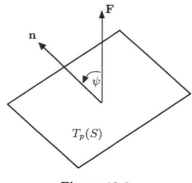

Figure 12.3.

From this analysis we expect the volume of the solid one unit high on the positive side of the surface to equal the total surface area and to be obtained by integrating the vector field $\mathbf{n}$ over $\mathbf{S}$. Indeed, we have

$$\iint\limits_{(S,\mathbf{n})} \mathbf{n} = \iint\limits_{U} (\mathbf{n} \cdot \mathbf{n}) \|\phi_u \times \phi_v\| dudv = \text{Surface area } (S).$$

It is now natural to define the surface integral of a *scalar-valued function* f on $(S, \mathbf{n})$ by identifying f with $f\mathbf{n}$, i.e.

$$\iint\limits_{(S,\mathbf{n})} f \overset{\text{def}}{=} \iint\limits_{(S,\mathbf{n})} f\mathbf{n} = \iint\limits_{S} f\big(\phi(u,v)\big) \|\phi_u \times \phi_v\| dudv$$

where ϕ is a parametrization consistent with the orientation. This is analogous to the way we defined, in Chapter 7, the integral of a scalar-valued function f along a curve by identifying f with fT. Although we needed an orientation to define this integral it is easily seen that

$$\iint\limits_{(S,\mathbf{n})} f = \iint\limits_{(S,-\mathbf{n})} f$$

and the value of the integral is independent of the orientation. This might suggest that it is possible to integrate a scalar-valued function over an *arbitrary* surface (see Definition 10.3). This is not true. What is required is an *orientable surface*, i.e. a surface which can be oriented. Simple surfaces are orientable. This means that we can use *any* parametrization to integrate a scalar-valued function over a simple surface.

    If the vector field $\mathbf{F}$ is represented by vectors emanating from the surface and if these all lie on the *same* side (of the surface) as $\mathbf{n}$ then the integral of $\mathbf{F}$

over $(S, \mathbf{n})$ is non-negative. On the right-hand diagram in Figure 12.2 we see that $\{\phi_u, \phi_v, \mathbf{n}\}$ follows the *right-hand rule* and note that $\mathbf{F}$ and $\mathbf{n}$ lie on the same side of $(S, \mathbf{n})$ if and only if the angle between them lies in $[-\pi/2, \pi/2]$.

Example 12.1

We evaluate $\iint\limits_{(S,\mathbf{n})} \mathbf{F}$ where $\mathbf{S}$ is the sphere of radius r with centre at the origin oriented outwards and $\mathbf{F}(x, y, z) = (x, y, z)$. Although a sphere is not a simple surface we can treat it as one for integration theory (see our remarks in the previous chapter). Here, however, we do not need any parametrization. The unit normals at a point P on S are $\pm P/\|P\|$ and since S is oriented outwards

$$\mathbf{n}(P) = \frac{P}{\|P\|} = \frac{P}{r}.$$

Hence

$$\mathbf{F}(P) \cdot \mathbf{n}(P) = P \cdot \frac{P}{r} = \frac{\|P\|^2}{r} = \frac{r^2}{r} = r$$

and

$$\iint\limits_{\mathbf{S}} \mathbf{F} = \iint\limits_{S} r \, dA = r(\text{Surface Area of } S) = r \cdot 4\pi r^2 = 4\pi r^3.$$

We defined vector-valued integrals by approximating small portions of the surface by appropriate parts of tangent planes and it should not come as a surprise that an orientation of the surface induces an orientation on each tangent plane. In oriented planes we can define *positive* and *negative area* and an anticlockwise sense of direction. This leads to another approach to vector-valued integration using *projections* onto the *coordinate planes* in $\mathbb{R}^3$.

Let $T_P(S)$ denote the tangent plane at the point P in the oriented surface $(S, \mathbf{n})$. The plane $T_P(S)$ can be regarded in its own right as a surface in $\mathbb{R}^3$ and the two unit normals at every point of $T_P(S)$ are $\pm \mathbf{n}(P)$. The oriented (tangent) plane $T_P(\mathbf{S}) = (T_P(S), \mathbf{n}(P))$ is said to have the orientation induced from $(S, \mathbf{n})$.

If $\mathbf{a}$ and $\mathbf{b}$ are non-zero tangent vectors in $T_P(\mathbf{S})$ we define the *angle* $\angle(\mathbf{a}, \mathbf{b})$ between $\mathbf{a}$ and $\mathbf{b}$ as the angle obtained by rotating $\mathbf{a}$ in an *anticlockwise direction* about $\mathbf{n}(P)$ until it points in the direction $\mathbf{b}$. This means that if $\angle(\mathbf{a}, \mathbf{b}) \neq 0$ then $\{\mathbf{a}, \mathbf{b}, \mathbf{n}(P)\}$ follows the *right-hand rule*. Order is important and

$$\angle(\mathbf{a}, \mathbf{b}) + \angle(\mathbf{b}, \mathbf{a}) = 2\pi.$$

To convince yourself that the choice of normal does make a difference you might like to try the following. Take a blank sheet of paper and label the two sides

A and B. Imagine that each side is oriented with normal pointing outwards. Take a point P on side A and draw two non-zero non-parallel vectors $\mathbf{a}$ and $\mathbf{b}$ originating at P. The angle θ_1 between $\mathbf{a}$ and $\mathbf{b}$ on side A is obtained by rotating to the left as in Figure 12.4. Now turn to side B. Using the transparency of the paper it is possible to use the sketch on side A to draw P, $\mathbf{a}$ and $\mathbf{b}$ on side B. If we again rotate $\mathbf{a}$ to the left to lie on $\mathbf{b}$ we get an angle θ_2 between $\mathbf{a}$ and $\mathbf{b}$ on side B (Figure 12.5). Note that $\theta_1 \neq \theta_2$ and, in fact, $\theta_1 + \theta_2 = 2\pi$.

Let $\pi(\mathbf{a}, \mathbf{b})$ denote the *parallelogram* spanned by the tangent vectors $\mathbf{a}$ and $\mathbf{b}$ at P, i.e. the figure obtained by placing end to end the vectors $\{\mathbf{a}, \mathbf{b}, -\mathbf{a}, -\mathbf{b}\}$. We define the *area of the parallelogram* $A(\mathbf{a}, \mathbf{b})$ to be

$$\|\mathbf{a}\| \cdot \|\mathbf{b}\| \sin\big(\angle(\mathbf{a}, \mathbf{b})\big) = \langle \mathbf{a} \times \mathbf{b}, \mathbf{n} \rangle .$$

Once more order is important and

$$A(\mathbf{a}, \mathbf{b}) = -A(\mathbf{b}, \mathbf{a}) .$$

In the previous chapter we calculated *surface area* using the *absolute value* of this area. Clearly

$$\sin\big(\angle(\mathbf{a}, \mathbf{b})\big) > 0 \iff 0 < \angle(\mathbf{a}, \mathbf{b}) < \pi \iff A\big(\pi(\mathbf{a}, \mathbf{b})\big) > 0 .$$

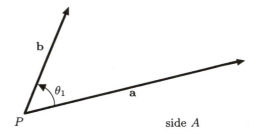

side A

Figure 12.4.

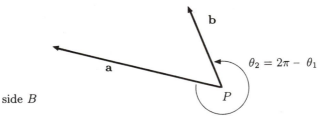

side B

Figure 12.5.

To display this we must choose a basis for $T_P(\mathbf{S})$. We say that an orthonormal basis $\{\mathbf{v}, \mathbf{w}\}$ is *positively ordered* if any of the following equivalent conditions are satisfied

$$\angle(\mathbf{v}, \mathbf{w}) = +\pi/2, \ \mathbf{v} \times \mathbf{w} = \mathbf{n}(P), \ \{\mathbf{v}, \mathbf{w}, \mathbf{n}(P)\} \text{ follows the right-hand rule.}$$

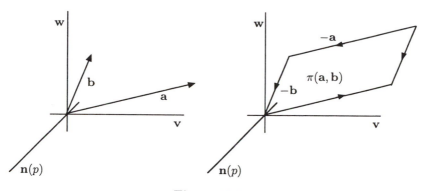

Figure 12.6.

If $\mathbf{a}$ and $\mathbf{b}$ are non-zero, non-parallel vectors in $T_P(\mathbf{S})$ then from Figure 12.6 we have

$$A\big(\pi(\mathbf{a}, \mathbf{b})\big) > 0 \iff \pi(\mathbf{a}, \mathbf{b}) \text{ is oriented in an } \textit{anticlockwise} \text{ direction.}$$

Moreover, if $\mathbf{a} = a_1\mathbf{v} + a_2\mathbf{w}$ and $\mathbf{b} = b_1\mathbf{v} + b_2\mathbf{w}$ then

$$A\big(\pi(\mathbf{a}, \mathbf{b})\big) = \det \begin{pmatrix} a_1 & a_2 \\ b_1 & b_2 \end{pmatrix} = a_1 b_2 - a_2 b_1 \, .$$

If $\varGamma$ is a piecewise smooth directed curve in $T_P(\mathbf{S})$ we define the *inside* of $\varGamma$ to be that part of $T_P(\mathbf{S})$ that lies on the *left* as we move along $\varGamma$ in the *positive* direction. It is possible (Figure 12.7) to approximate the inside of $\varGamma$ by oriented rectangles and to define the area of the inside as the limit of the sum of the areas of these rectangles and, in this way, we see that the area of the inside of an anticlockwise oriented curve in the $\{\mathbf{v}, \mathbf{w}\}$ plane is positive.

Our aim now is to use parametrizations to transfer integration from an oriented surface in $\mathbb{R}^3$ to integration over subsets of the *coordinate planes* in $\mathbb{R}^3$, i.e. the xy-plane, the yz-plane and the xz-plane. Each of these planes is a surface in $\mathbb{R}^3$ and to define our integral we must assign a positive side to each of these – for the same reason that we required positive intervals in $\mathbb{R}$.

We first define the *positive unit vectors* in the x, y and z directions in $\mathbb{R}^3$ as $(1, 0, 0)$, $(0, 1, 0)$ and $(0, 0, 1)$. These choices are natural in view of the way we

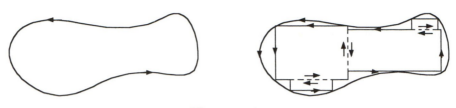

Figure 12.7.

sketch graphs in $\mathbb{R}^2$ and $\mathbb{R}^3$. Consider now the xy-plane in $\mathbb{R}^3$ as the surface defined by $f^{-1}(0)$ where $f(x, y, z) = z$. The unit normals to this surface are $\pm(0, 0, 1)$ and we must choose between them to define a positive side to the xy-plane. We use the positive unit vectors in the x and y directions *in that order* and the *right-hand rule* or *cross product* to define the *positive side* of the xy-plane as that oriented by the normal

$$(1, 0, 0) \times (0, 1, 0) = (0, 0, 1).$$

We denote the positively oriented xy-plane in $\mathbb{R}^3$ by $\mathbb{R}^2_{(x,y)}$. Similarly the positive side of the yz-plane is defined by the normal $(0, 1, 0) \times (0, 0, 1) = (1, 0, 0)$ and denoted by $\mathbb{R}^2_{(y,z)}$ and the positive side of the zx-plane is defined by $(0, 0, 1) \times (1, 0, 0) = (0, 1, 0)$ and denoted by $\mathbb{R}^2_{(z,x)}$.

In Figure 12.8 we see the direction of rotation (i.e. rotation to the *left*) used to measure angles in each of the coordinate planes and also the direction of a closed anticlockwise oriented curve.

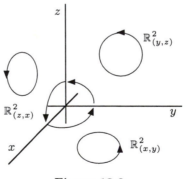

Figure 12.8.

If we look at each individual coordinate plane we get, using Figure 12.8, the diagrams in Figure 12.9 showing the anticlockwise directions.

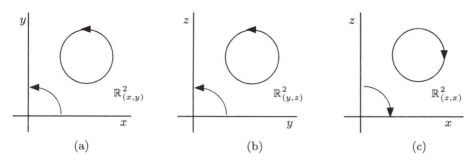

Figure 12.9.

In Chapter 9 we discussed Green's theorem

$$\oint_\Gamma Pdx + Qdy = \iint_\Omega \left(\frac{\partial Q}{\partial x} - \frac{\partial P}{\partial y}\right)dxdy \tag{12.1}$$

where the boundary of Ω, Γ is oriented in an *anticlockwise* direction. From (12.1) it does not appear that Ω has been assigned any orientation but, in fact, if we identify $\mathbb{R}^2$ with $\mathbb{R}^2_{(x,y)} \subset \mathbb{R}^3$ then the integration on $\mathbb{R}^2$ is really over the *positive side* of $\mathbb{R}^2_{(x,y)}$. This is the matching up of orientations in Green's theorem.

Let $\mathbf{F} = (f, g, h)$ denote a continuous vector field on the simple oriented surface $\mathbf{S}$. Let

$$\phi: (u, v) \in U \longrightarrow \big(x(u, v), y(u, v), z(u, v)\big)$$

denote a parametrization consistent with the orientation. We now take an independent approach to defining separately $\iint_S (f, 0, 0)$, $\iint_S (0, g, 0)$ and $\iint_S (0, 0, h)$.

First observe that $(f, 0, 0)$ is parallel to the x-axis and may be represented as in Figure 12.10, where we have rotated the axes but preserved the orientation.

Since f points in the direction of the normal $(1, 0, 0)$ to $\mathbb{R}^2_{(y,z)}$ it is natural to project onto $\mathbb{R}^2_{(y,z)}$ and to define $\iint_S (f, 0, 0)$ as $\iint_{P_{y,z}(\mathbf{S}) \subset \mathbb{R}^2_{(y,z)}} f$ where $P_{y,z}$ is the projection $(x, y, z) \to (y, z)$ and $P_{y,z}(\mathbf{S})$ is given the induced orientation from $\mathbb{R}^2_{(y,z)}$. Although the set $P_{y,z}(\mathbf{S})$ may not be open and the mapping

$$P_{y,z}: (u, v) \longrightarrow \big(y(u, v), z(u, v)\big)$$

is not necessarily a parametrization of $P_{y,z}(\mathbf{S})$ we *may* proceed as if they were (in some cases a knowledge of the shape or geometry of $P_{y,z}(\mathbf{S})$ may suggest a different "parametrization"). Take a partition of U into rectangles, $(R_{ij})_{ij}$,

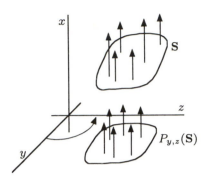

Figure 12.10.

transfer the partition to $P_{y,z}(\mathbf{S})$ by using $P_{y,z}$, form a Riemann sum of f with respect to this partition, i.e.

$$\sum_i \sum_j f\big(\phi(u_i, v_j)\big) \times \Big[\text{Area } \big(P_{y,z}(R_{ij})\big) \text{ in } \mathbb{R}^2_{(y,z)} \Big]$$

$$\approx \sum_i \sum_j f\big(\phi(u_i, v_j)\big) \times \Big[A\big(\pi\big((y_u, z_u), (y_v, z_v)\big)\big) \Big] (u_i, v_j) \Delta u_i \Delta v_j$$

$$= \sum_i \sum_j f\big(\phi(u_i, v_j)\big) \begin{vmatrix} y_u & z_u \\ y_v & z_v \end{vmatrix} (u_i, v_j) \Delta u_i \Delta v_j \, .$$

On taking a limit we get

$$\iint_{\mathbf{S}} (f, 0, 0) = \iint_U f\big(\phi(u, v)\big) \begin{vmatrix} y_u & z_u \\ y_v & z_v \end{vmatrix} du\, dv \, .$$

We proceed in the same way with $(0, g, 0)$ and obtain the diagram shown in Figure 12.11, which implies that z precedes x (see also Figure 12.9(c)) and so the correctly oriented "parametrization" in this case is

$$(u, v) \longrightarrow \big(z(u, v), x(u, v)\big) \, .$$

This leads to the definition

$$\iint_{\mathbf{S}} (0, g, 0) = \iint_U g\big(\phi(u, v)\big) \begin{vmatrix} z_u & x_u \\ z_v & x_v \end{vmatrix} du\, dv \, .$$

Similarly we are led to

$$\iint_{\mathbf{S}} (0, 0, h) = \iint_U h\big(\phi(u, v)\big) \begin{vmatrix} x_u & y_u \\ x_v & y_v \end{vmatrix} du\, dv \, .$$

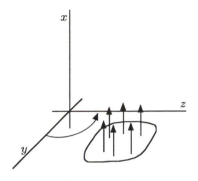

Figure 12.11.

Although each of these integrals looks rather complicated when we add them together we obtain a familiar expression. Since

$$\phi(u,v) = \big(x(u,v), y(u,v), z(u,v)\big)$$

we have $\phi_u = (x_u, y_u, z_u)$ and $\phi_v = (x_v, y_v, z_v)$. Hence

$$\phi_u \times \phi_v = \begin{vmatrix} \mathbf{i} & \mathbf{j} & \mathbf{k} \\ x_u & y_u & z_u \\ x_v & y_v & z_v \end{vmatrix} = \left(\begin{vmatrix} y_u & z_u \\ y_v & z_v \end{vmatrix}, -\begin{vmatrix} x_u & z_u \\ x_v & z_v \end{vmatrix}, \begin{vmatrix} x_u & y_u \\ x_v & y_v \end{vmatrix} \right)$$

$$\mathbf{F} \cdot \phi_u \times \phi_v = f\begin{vmatrix} y_u & z_u \\ y_v & z_v \end{vmatrix} - g\begin{vmatrix} x_u & z_u \\ x_v & z_v \end{vmatrix} + h\begin{vmatrix} x_u & y_u \\ x_v & y_v \end{vmatrix}$$

$$= \begin{vmatrix} f & g & h \\ x_u & y_u & z_u \\ x_v & y_v & z_v \end{vmatrix} = \det\begin{pmatrix} \mathbf{F} \\ \phi_u \\ \phi_v \end{pmatrix}$$

and

$$\iint_S (f,0,0) + \iint_S (0,g,0) + \iint_S (0,0,h)$$

$$= \iint_U \left(f\begin{vmatrix} y_u & z_u \\ y_v & z_v \end{vmatrix} + g\begin{vmatrix} z_u & x_u \\ z_v & x_v \end{vmatrix} + h\begin{vmatrix} x_u & y_u \\ x_v & y_v \end{vmatrix} \right) dudv \qquad (12.2)$$

$$= \iint_U \mathbf{F}\big(\phi(u,v)\big) \cdot \phi_u \times \phi_v(u,v)\, dudv \qquad (12.3)$$

$$= \iint_U \det\begin{pmatrix} \mathbf{F} \\ \phi_u \\ \phi_v \end{pmatrix} dudv. \qquad (12.4)$$

Formula (12.3) is our original definition of the integral of $\mathbf{F}$ over $\mathbf{S}$ and thus (12.2) and (12.4) are new expressions for $\iint_S \mathbf{F}$. From (12.3) and the results of the previous chapter we also have

$$
\iint_S \mathbf{F} = \iint_U (\mathbf{F} \cdot \mathbf{n}) \|\phi_u \times \phi_v\| du dv
$$

$$
= \iint_U (\mathbf{F} \cdot \mathbf{n}) \sqrt{EG - F^2} du dv . \tag{12.5}
$$

Using the notation

$$
\frac{\partial(x,y)}{\partial(u,v)} = \begin{vmatrix} x_u & y_u \\ x_v & y_v \end{vmatrix}, \quad \text{etc.}
$$

we obtain yet another formula for the integral:

$$
\iint_S \mathbf{F} = \iint_U (\mathbf{F} \cdot \mathbf{n}) \left[\left(\frac{\partial(x,y)}{\partial(u,v)} \right)^2 + \left(\frac{\partial(y,z)}{\partial(u,v)} \right)^2 + \left(\frac{\partial(z,x)}{\partial(u,v)} \right)^2 \right]^{1/2} du dv . \tag{12.6}
$$

In view of formula (12.5) the notation $\iint_S \langle \mathbf{F}, \mathbf{n} \rangle dA$ is also used in place of $\iint_{(S,\mathbf{n})} \mathbf{F}$ where dA denotes surface area.

We have established five formulae for calculating surface integrals. Furthermore, our excursion into oriented planes has led to geometric insights on the construction of the integral and to a method of evaluating integrals by projecting onto the coordinate planes. There are still a number of topics to be sorted out, e.g. independence of the parametrization, and we discuss these in examples as we proceed.

Example 12.2

In this simple example we use projections to calculate the area of the triangle with vertices $(1,0,0)$, $(0,1,0)$ and $(0,0,1)$. The fact that we can easily find the answer independently allows us to check our solution.

From Figure 12.12, $P = (1/2, 1/2, 0)$ and

$$
\|(1/2, 1/2, 0) - (0,0,1)\| = \left(\frac{1}{4} + \frac{1}{4} + 1 \right)^{1/2} = \sqrt{\frac{3}{2}} .
$$

Since $\|(1,0,0) - (0,1,0)\| = \sqrt{2}$ the (surface) area is $1/2 \cdot \sqrt{2} \cdot \sqrt{3/2} = \sqrt{3}/2$. We now calculate the area using $\iint_{(S,\mathbf{n})} \mathbf{n}$. Since the triangle is part of a *plane* the normal $\mathbf{n}$ is constant on $\mathbf{S}$. By symmetry it is easily seen that the triangle lies in the plane $x + y + z = 1$ and hence the unit normals are $\pm(1/\sqrt{3}, 1/\sqrt{3}, 1/\sqrt{3})$. Let us take $(1/\sqrt{3}, 1/\sqrt{3}, 1/\sqrt{3})$ as our unit normal. By symmetry we only have to

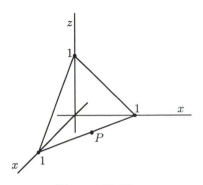

Figure 12.12.

calculate $\iint\limits_{(S,\mathbf{n})} (1/\sqrt{3},0,0)$. Clearly $P_{y,z}$ projects onto the triangle $A_1 \subset \mathbb{R}^2_{(y,z)}$
with vertices $(0,0)$, $(0,1)$ and $(1,0)$ and

$$\iint\limits_{(S,\mathbf{n})} (\frac{1}{\sqrt{3}},0,0) = \frac{1}{\sqrt{3}} \text{ Area } (A_1) = \frac{1}{\sqrt{3}} \cdot \frac{1}{2}.$$

Hence

$$\iint\limits_{(S,\mathbf{n})} \mathbf{n} = 3 \cdot \frac{1}{\sqrt{3}} \cdot \frac{1}{2} = \frac{\sqrt{3}}{2}$$

and this agrees with our earlier result.

Example 12.3

We compute

$$\iint\limits_{\mathbf{S}} \frac{(x,y,z-a)}{\sqrt{2az - z^2}}$$

where $\mathbf{S}$ is the part of the surface $x^2 + y^2 + (z-a)^2 = a^2$ which lies inside the
cylinder $x^2 + y^2 = ay$ and underneath the plane $z = a$ oriented with outward
pointing normal.

We first sketch the surface over which we are integrating. Initially this
appears as a rather formidable task but by sketching each part separately and
then combining them it becomes relatively simple. The surface $x^2 + y^2 + (z -
a)^2 = a^2$ is a sphere of radius a with centre $(0,0,a)$. The equation $x^2 + y^2 = ay$
can be rewritten as

$$x^2 + y^2 - ay - \frac{a^2}{4} = \frac{a^2}{4}$$

i.e.

$$x^2 + \left(y - \frac{a}{2}\right)^2 = \left(\frac{a}{2}\right)^2$$

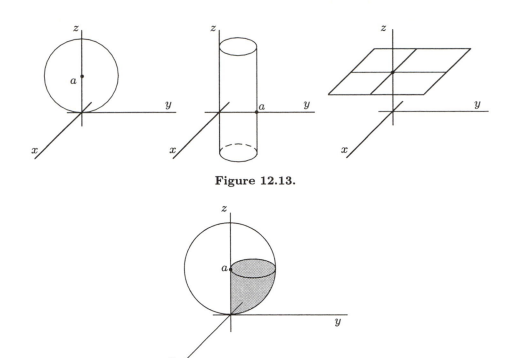

Figure 12.13.

Figure 12.14.

and this surface is the cylinder parallel to the z-axis above the circle with centre $(0, a/2)$ and radius $a/2$ in the xy-plane. The plane $z = a$ is parallel to the xy-plane and a units above it. We sketch each of these separately in Figure 12.13 and combine them in Figure 12.14.

We thus are integrating over the lower right-hand portion of the sphere and use geographical coordinates about the point $(0, 0, a)$

$$F: (\theta, \psi) \longrightarrow (a \cos\theta \cos\psi, a \cos\theta \sin\psi, a + a \sin\theta).$$

Since θ refers to latitude and we are considering the lower portion we have $-\pi/2 < \theta < 0$. From the sketch $y > 0$. Since $\cos\theta > 0$, we have $\sin\psi > 0$, i.e. $0 < \psi < \pi$. The outward normal to the sphere of radius a with centre $(0, 0, 0)$ at (x, y, z) is $(x, y, z - a)/a$. Hence

$$\iint\limits_{S} \frac{(x, y, z - a)}{\sqrt{2az - z^2}} = \iint\limits_{S} \frac{(x, y, z - a)}{\sqrt{2az - z^2}} \cdot \frac{(x, y, z - a)}{a} dA$$

$$= \iint\limits_{S} \frac{a}{\sqrt{2az - z^2}} dA.$$

Now

$$2az - z^2 = -a^2 + 2az - z^2 + a^2 = a^2 - (a - z)^2$$

and changing this into our new coordinates we get

$$\sqrt{2az - z^2} = (a^2 - a^2 \sin^2 \theta)^{1/2} = (a^2 \cos^2 \theta)^{1/2} = a \cos \theta\,.$$

From Example 10.3 we recall that

$$dA = \sqrt{EG - F^2}\,d\theta d\psi = a^2 \cos \theta d\theta d\psi\,.$$

Hence

$$\iint_S \frac{(x, y, z - a)}{\sqrt{2az - z^2}} = \int_{-\pi/2}^{0} \int_{0}^{\pi} \frac{a^2 \cos \theta \cdot a}{a \cos \theta}\,d\theta d\psi = a^2(\pi/2)\pi = \pi^2 a^2/2\,.$$

EXERCISES

12.1 Find $\iint_S \mathbf{F}$ where $\mathbf{F}(x, y, z) = (1, 2, 3)$ and $\mathbf{S}$ is the triangle with vertices $(1, 0, 0)$, $(0, 2, 0)$, $(0, 0, 3)$ oriented so that the origin is on the negative side.

12.2 By using projections and the portion of the sphere that lies in the first octant calculate the area of a sphere of radius r.

12.3 Express as an integral over a region in $\mathbb{R}^2$ the integral $\iint_S \mathbf{F}$ where $\mathbf{F}(x, y, z) = \left(-2/y^3, -6xy^2, 2z^3/x^3\right)$, $\mathbf{S}$ is the graph of $f(u, v) = uv^3$, $(u, v) \in [1, 2] \times [1, 2]$ and the parametrization $\phi(u, v) = (u, v, uv^3)$ is consistent with the orientation. Evaluate the integral.

12.4 Let $0 < a < b$ and let Γ denote the circle of centre $(b, 0)$ and radius a in the xz-plane. Let S denote the surface obtained by rotating Γ about the z-axis. If S is oriented with outward normal and

$$\mathbf{F}(x, y, z) = \left(x - \frac{bx}{\sqrt{x^2 + y^2}}, y - \frac{by}{\sqrt{x^2 + y^2}}, z\right)$$

show that

$$\iint_S \mathbf{F} = 4\pi^2 a^2 b\,.$$

12.5 Evaluate

(a) $\displaystyle\iint_S y^2 + z^2$

(b) $\displaystyle\iint_S \frac{1}{\left(x^2 + y^2 + (z+a)^2\right)^{1/2}}$

where $\mathbf{S}$ is the portion of the sphere $x^2 + y^2 + z^2 = a^2$ above the xy-plane oriented by the outward normal.

12.6 Describe the surface $S = \{(x, y, z) : z^2 = x^2 + y^2, 1 \le z \le 3\}$. If $\mathbf{n}$ is the outward pointing normal to S find

$$\iint_{(S,\mathbf{n})} \frac{(-xz, -yz, x^2 + y^2)}{x^2 + y^2}.$$

12.7 Let S denote the portion of the level set $z = \tan^{-1}(y/x)$ which lies between the planes $z = 0$ and $z = 2\pi$, inside the cone $x^2 + y^2 = z^2$ and outside the cylinder $x^2 + y^2 = \pi^2$. Let

$$\mathbf{G}(x, y, z) = \frac{(-x, -y, -z)}{(x^2 + y^2 + z^2)^{3/2}}.$$

Show that

$$F(r, \theta) = (r\cos\theta, r\sin\theta, \theta), \quad \pi < r < \theta, \ \pi < \theta < 2\pi$$

is a parametrization of S. If $\mathbf{n} = \dfrac{F_r \times F_\theta}{\|F_r \times F_\theta\|}$ evaluate $\displaystyle\iint_{(S,\mathbf{n})} \mathbf{G}$.

13
Stokes' Theorem

Summary. *We discuss Stokes' theorem for oriented surfaces in* $\mathbb{R}^3$.

Stokes' theorem, the *fundamental theorem of calculus for surfaces*, generalises Green's theorem to oriented surfaces $\mathbf{S} = (S, \mathbf{n})$ with *edge* or *boundary* Γ (the term edge avoids confusion with our other use of the word boundary) consisting of a finite number of piecewise smooth directed curves. We suppose that the *positive side* of $\mathbf{S}$ lies on the *left-hand side* as we move along Γ in the *positive direction*. In practice this consistency between the orientations of the surface and its edge may be verified by sketching. In certain cases the normal $\mathbf{n}$ admits a *continuous extension* to the boundary and a parametrization P of the boundary has the correct orientation if at one point, say $P(t_0)$, we have $P'(t_0) \cdot \mathbf{n}(P(t_0)) > 0$ where $\mathbf{n}(P(t_0))$ is the value of the extension of $\mathbf{n}$ at $P(t_0)$. If a consistent parametrization of the surface extends to give a parametrization of the boundary then the boundary is also correctly directed (see Example 13.4).

Theorem 13.1 (Stokes' Theorem)

Let $\mathbf{S} = (S, \mathbf{n})$ denote an oriented surface in $\mathbb{R}^3$ with boundary Γ consisting of a finite number of piecewise smooth directed curves. We suppose that the positive side of $\mathbf{S}$ lies on the left of the positive side of Γ. If $\mathbf{F}$ is a smooth vector field on $S \cup \Gamma$ then

$$\int_\Gamma \mathbf{F} = \iint_\mathbf{S} \mathrm{curl}(\mathbf{F}) \tag{13.1}$$

153

i.e.

$$\int_\Gamma \langle \mathbf{F}, T \rangle ds = \iint_S \langle \mathrm{curl}(\mathbf{F}), \mathbf{n} \rangle dA$$

where T is the unit tangent to the directed curves Γ.

The proof, which we omit, is obtained by applying Green's theorem to the projections onto the coordinate planes. In Chapters 6 and 12 we developed techniques to evaluate the left- and right-hand sides of (12.1), respectively. Thus the only new factor in Stokes' theorem is the correlation between the orientations of the surface and its boundary.

Example 13.1

We use Stokes' theorem to evaluate the line integral

$$\int_C -y^3 dx + x^3 dy - z^3 dz$$

where C is the intersection of the cylinder $x^2 + y^2 = 3$ and the plane $x + y + z = 1$ and the orientation on C is anticlockwise when viewed from a point sufficiently high up on the z-axis. Let $\mathbf{S}$ denote the portion of the plane inside the cylinder oriented so that the normal lies above the surface (Figure 13.1).

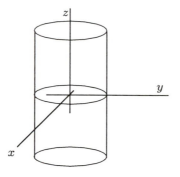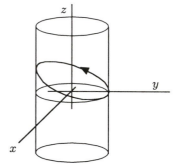

Figure 13.1.

Let $\mathbf{F} = (-y^3, x^3, -z^3)$ then

$$\mathrm{curl}(\mathbf{F}) = \begin{vmatrix} \mathbf{i} & \mathbf{j} & \mathbf{k} \\ \dfrac{\partial}{\partial x} & \dfrac{\partial}{\partial y} & \dfrac{\partial}{\partial z} \\ -y^3 & x^3 & -z^3 \end{vmatrix} = (0, 0, 3x^2 + 3y^2).$$

By Stokes' theorem

$$\int_C \mathbf{F} = \iint_S \text{curl}(\mathbf{F}) = \iint_S (0, 0, 3x^2 + 3y^2).$$

Since only the final coordinate of $\text{curl}(\mathbf{F})$ is non-zero our analysis in the previous chapter implies that we only need consider the projection of $\mathbf{S}$ onto $\mathbb{R}^2_{(x,y)}$. As C projects onto an anticlockwise oriented curve Γ_1 (Figure 13.1) our projection is onto the positive side of $\mathbb{R}^2_{(x,y)}$. Hence

$$\int_C \mathbf{F} = \iint_{x^2+y^2 \le 3} (3x^2 + 3y^2) dx dy.$$

We parametrise the surface $x^2 + y^2 < 3$ in $\mathbb{R}^3$ by

$$\phi(r, \theta) = (r\cos\theta, r\sin\theta, 0), \quad 0 < r < \sqrt{3},\ 0 < \theta < 2\pi.$$

We have $\phi_r = (\cos\theta, \sin\theta, 0)$ and $\phi_\theta = (-r\sin\theta, r\cos\theta, 0)$. Hence

$$\phi_r \times \phi_\theta = \begin{vmatrix} \mathbf{i} & \mathbf{j} & \mathbf{k} \\ \cos\theta & \sin\theta & 0 \\ r\sin\theta & r\cos\theta & 0 \end{vmatrix} = (0, 0, r).$$

Since

$$\frac{\phi_r \times \phi_\theta}{\|\phi_r \times \phi_\theta\|} = (0, 0, 1)$$

our parametrization is consistent with the positive orientation of $\mathbb{R}^2_{(x,y)}$. We have

$$\int_C \mathbf{F} = \int_0^{\sqrt{3}} \int_0^{2\pi} 3r^2 \cdot \|\phi_r \times \phi_\theta\| dr d\theta = 6\pi \int_0^{\sqrt{3}} r^3 dr = \frac{27\pi}{2}.$$

Moral: A reasonable sketch is not just optional but necessary. The form of $\mathbf{F}$, i.e. the fact that the first two coordinates were zero, combined with information on how surface integrals can be projected onto the coordinate planes greatly simplified the calculations required.

Example 13.2

We evaluate $\iint_S \text{curl}\,(\mathbf{F})$ where

$$\mathbf{F}(x, y, z) = (y^2 \cos xz, x^3 e^{yz}, -e^{-xyz})$$

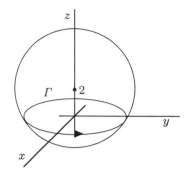

Figure 13.2.

and $\mathbf{S}$ is the portion of the sphere $x^2 + y^2 + (z-2)^2 = 8$ which lies above the xy-plane oriented outwards. The edge or boundary of $\mathbf{S}$, Γ, is where the sphere cuts the xy-plane, i.e. where $z = 0$. We have $x^2 + y^2 + 4 = 8$, i.e. $x^2 + y^2 = 4$ (Figure 13.2).

Since the positive side of the sphere is the outside we see from Figure 13.2 that the surface $\mathbf{S}$ is on the left as we move along Γ in an anticlockwise direction in the xy-plane. This gives us our direction along Γ. By Stokes' theorem

$$\int_{\Gamma} \mathbf{F} = \iint_{\mathbf{S}} \mathrm{curl}(\mathbf{F}).$$

But Γ with anticlockwise direction is also the boundary or edge of the surface

$$\mathbf{S}_1 = \{(x, y, z) : x^2 + y^2 < 4, z = 0\}$$

oriented by the normal $(0, 0, 1)$. Hence a further application of Stokes' theorem implies

$$\int_{\Gamma} \mathbf{F} = \iint_{\mathbf{S}_1} \mathrm{curl}(\mathbf{F}).$$

Now

$$\mathrm{curl}(\mathbf{F}) = \begin{vmatrix} \mathbf{i} & \mathbf{j} & \mathbf{k} \\ \dfrac{\partial}{\partial x} & \dfrac{\partial}{\partial y} & \dfrac{\partial}{\partial z} \\ y^2 \cos xz & x^3 e^{yz} & -e^{-xyz} \end{vmatrix}.$$

Since, however, $\mathbf{S}_1$ projects onto smooth curves, which have zero surface area in $\mathbb{R}^2_{(y,z)}$ and $\mathbb{R}^2_{(x,z)}$, it suffices to consider the final coordinate of curl $(\mathbf{F})$. This is

$$\frac{\partial}{\partial x}\left(x^3 e^{yz}\right) - \frac{\partial}{\partial y}\left(y^2 \cos xz\right) = 3x^2 e^{yz} - 2y \cos xz$$

and, since $z = 0$ on $\mathbf{S}_1$, we need only evaluate

$$\iint\limits_{x^2+y^2\leq 4} (3x^2 - 2y)dxdy\,.$$

By symmetry

$$\iint\limits_{x^2+y^2\leq 4} (-2y)dxdy = 0\,.$$

If we use the parametrization

$$(r, \theta) \longrightarrow (r\cos\theta, r\sin\theta, 0), \quad 0 < r < 2,\ 0 < \theta < 2\pi$$

then

$$\iint\limits_{\mathbf{S}_1} \mathbf{F} = \int_0^2 \int_0^{2\pi} 3r^2\cos^2\theta r\,dr\,d\theta = \int_0^{2\pi}\cos^2\theta d\theta \int_0^2 3r^3\,dr = 12\pi\,.$$

Moral: A closed curve may be the edge or boundary of more than one surface and a suitable choice (of surface) may simplify calculations. Projections and symmetry are helpful.

Example 13.3

We wish to use Stokes' theorem to find a suitable orientation of the curve of intersection of $x^2 + y^2 + z^2 = a^2$ and $x + y + z = 0$, Γ, so that

$$\int_\Gamma ydx + zdy + xdz = \sqrt{3}\pi a^2\,.$$

The curve Γ is the intersection of a sphere and a plane through the centre of the sphere and hence is a "great circle" or "equator" on a sphere (see Figure 13.3).

The curve Γ is the edge or boundary of the two hemispheres on either side of it and also of a portion of the plane $x + y + z = 0$. Which we use will depend on the function being integrated. Let $\mathbf{F}(x, y, z) = (y, z, x)$. We have

$$\mathrm{curl}(\mathbf{F}) = \begin{vmatrix} \mathbf{i} & \mathbf{j} & \mathbf{k} \\ \dfrac{\partial}{\partial x} & \dfrac{\partial}{\partial y} & \dfrac{\partial}{\partial z} \\ y & z & x \end{vmatrix} = (-1, -1, -1)\,.$$

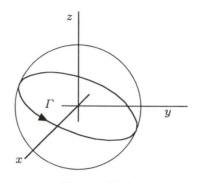

Figure 13.3.

The plane $x + y + z = 0$ has unit normals $\pm(1,1,1)/\sqrt{3} = \mathbf{n}_1(x,y,z)$ while the sphere has unit normals $\pm(x,y,z)/a = \mathbf{n}_2(x,y,z)$. By Stokes' theorem

$$\int_\Gamma y\,dx + z\,dy + x\,dz = \iint\limits_{(S,\mathbf{n})} (-1,-1,-1)$$

where S and $\mathbf{n}$ have to be chosen and Γ directed. If we take $\mathbf{S}$ as part of the plane with normal $(1,1,1)/\sqrt{3}$ then

$$\iint\limits_S (-1,-1,-1) = \iint\limits_S (-1,-1,-1) \cdot \frac{(1,1,1)}{\sqrt{3}} dA = -\frac{3}{\sqrt{3}} \iint\limits_S dA$$

$$= -\sqrt{3}\pi a^2$$

since S is a disc of radius a.

Since we obtained a negative answer we have been using the incorrect orientation on S and so take $(-1/\sqrt{3}, -1/\sqrt{3}, -1/\sqrt{3})$ as the normal which describes the orientation. Hence Γ is oriented as in Figure 13.3, i.e. it appears clockwise when looked at from, say, the point $(1,1,1)$ or from any point sufficiently far out in the first octant.

Moral: The curl of a vector field is a form of derivative. If the entries are linear, as in this example, the curl is constant. If the entries are of degree 2 then the curl has linear entries.

Example 13.4

In this example we verify Stokes' theorem for the portion S of the surface $z = \tan^{-1}(y/x)$ which lies inside the cone $x^2 + y^2 = z^2$ and between the planes $z = 0$ and $z = 2\pi$ by using the vector field

$$\mathbf{F}(x,y,z) = xz\mathbf{i} + yz\mathbf{j} - (x^2 + y^2)\mathbf{k}.$$

We first examine the surface $z = \tan^{-1}(y/x)$. This can be parametrised as a graph using Cartesian coordinates (see Example 10.1) but it is preferable to use polar coordinates.

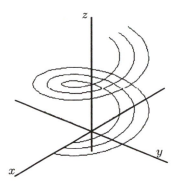

Figure 13.4.

Let $x = r \cos \theta$, $y = r \sin \theta$ then

$$\tan^{-1}\left(\frac{y}{x}\right) = \tan^{-1}\left(\frac{r \sin \theta}{r \cos \theta}\right) = \tan^{-1}(\tan \theta) = \theta$$

and we obtain the parametrization

$$(r, \theta) \longrightarrow (r \cos \theta, r \sin \theta, \theta) \tag{13.2}$$

where $r > 0$ and $0 < \theta < 2\pi$.

What sort of surface is this? Well, if we fix different values of r and let θ vary we obtain a *helix* (see Example 5.1). In Figure 13.4 we have sketched the surface for $1/2 \leq r \leq 1$. We are considering the portion of the surface in Figure 13.4, extended over all r, which lies inside the cone and between the two planes. It is thus a *screw-shaped* surface or *spiral staircase* with the spirals or steps getting wider as we rise. The fact that the surface lies between the planes $z = 0$ and $z = \pi$ means that we have just one full twist of the screw or one turn of the staircase (Figure 13.5).

We can use the parametrization (12.2) but the restriction caused by lying inside the other surfaces means that we must restrict the range. Translating the boundaries into polar coordinates, we get

$$x^2 + y^2 = r^2 \cos^2 \theta + r^2 \sin^2 \theta = r^2 = z^2$$

and since $0 \leq z \leq 2\pi$ this implies $r = z$, $0 \leq r \leq 2\pi$. Hence our parametrization of the surface is

$$f : (r, \theta) \longrightarrow (r \cos \theta, r \sin \theta, \theta), \quad 0 < r < \theta, \ 0 < \theta < 2\pi.$$

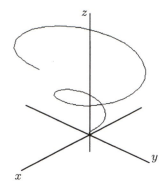

Figure 13.5.

We have $\mathbf{F}(f(r,\theta)) = (r\theta\cos\theta, r\theta\sin\theta, -r^2)$ and

$$\mathrm{curl}(\mathbf{F}) = \begin{vmatrix} \mathbf{i} & \mathbf{j} & \mathbf{k} \\ \dfrac{\partial}{\partial x} & \dfrac{\partial}{\partial y} & \dfrac{\partial}{\partial z} \\ xz & yz & -(x^2+y^2) \end{vmatrix} = -3y\mathbf{i} + 3x\mathbf{j} = (-3r\sin\theta, 3r\cos\theta, 0) .$$

We define our orientation on **S** by

$$f_r \times f_\theta = \begin{vmatrix} \mathbf{i} & \mathbf{j} & \mathbf{k} \\ \cos\theta & \sin\theta & 0 \\ -r\sin\theta & r\cos\theta & 1 \end{vmatrix} = (\sin\theta, -\cos\theta, r) .$$

Hence

$$\iint_S \mathrm{curl}(\mathbf{F}) = \int_0^{2\pi} \int_0^\theta \langle \mathrm{curl}(\mathbf{F}), f_r \times f_\theta \rangle dr d\theta$$
$$\phantom{\iint_S \mathrm{curl}(\mathbf{F})}_{(\theta)\ (r)}$$

$$= \int_0^{2\pi} \left(\int_0^\theta -3r dr \right) d\theta = \int_0^{2\pi} -\frac{3r^2}{2} \Big|_0^\theta d\theta$$

$$= -\frac{3}{2} \int_0^{2\pi} \theta^2 d\theta = -\frac{3}{2}\frac{\theta^3}{3} \Big|_0^{2\pi}$$

$$= -\frac{(2\pi)^3}{2} = -4\pi^3 .$$

The parametrization of S is over the set U in $\mathbb{R}^2$ given in Figure 13.6 and the boundary or edge of the surface can be found by examining

$$f(r,\theta) = (r\cos\theta, r\sin\theta, \theta) \tag{13.3}$$

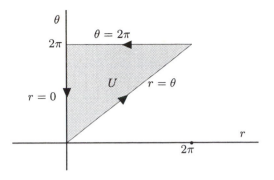

Figure 13.6.

on the boundary of U. We can also look at it geometrically by examining where the boundary curves intersect S. We first look at the curve of intersection of the cone and the screw. The cone is parametrised (Examples 10.2 and 10.3) by

$$(r, \theta) \longrightarrow (r \cos \theta, r \sin \theta, r)$$

and comparing this with (13.3) we get a curve of intersection Γ_1 when $r = \theta$. This curve is parametrised by

$$\theta \longrightarrow (\theta \cos \theta, \theta \sin \theta, \theta), \quad 0 \le \theta \le 2\pi.$$

The curve Γ_1 joins the origin to $P_1 = (2\pi, 0, 2\pi)$. The second curve Γ_2 is obtained by putting $\theta = 2\pi$ and from (13.3) this is parametrised by

$$r \longrightarrow (r \cos 2\pi, r \sin 2\pi, 2\pi) = (r, 0, 2\pi)$$

where $0 < r < 2\pi$. This is the straight line joining P_1 to $(0, 0, 2\pi)$. From Figure 13.6 it runs in the *negative direction* and so we must reverse the orientation.

The third curve Γ_3 is obtained by letting $r = 0$ in (12.3) and we have a parametrization

$$\theta \longrightarrow (0, 0, \theta), \quad 0 \le \theta \le 2\pi.$$

Γ_3 joins $(0, 0, 2\pi)$ to the origin and $\Gamma = \Gamma_1 \cup \Gamma_2 \cup \Gamma_3$ is a closed piecewise smooth directed curve (Figure 13.7).

We have oriented the surface and directed its boundary. Are these consistent in order to apply Stokes' theorem? Yes, because the parametrization f of Γ is obtained from a *continuous extension* of a consistent parametrization of the surface. You have already seen two other ways in which it is possible to check this consistency. We now have to evaluate

$$\int_\Gamma \mathbf{F} = \int_{\Gamma_1} \mathbf{F} + \int_{\Gamma_2} \mathbf{F} + \int_{\Gamma_3} \mathbf{F}.$$

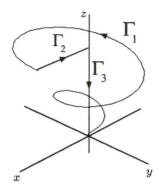

Figure 13.7.

We have

$$\int_{\Gamma_1} \mathbf{F} = \int_0^{2\pi} \left(\frac{d}{d\theta}(\theta \cos\theta, \theta \sin\theta, \theta) \right) \cdot (\theta^2 \cos\theta, \theta^2 \sin\theta, -\theta^2) \, d\theta$$

$$= \int_0^{2\pi} (\theta^2 \cos^2\theta - \theta^3 \sin\theta \cos\theta + \theta^2 \sin^2\theta + \theta^3 \sin\theta \cos\theta - \theta^2) \, d\theta$$

$$= \int_0^{2\pi} 0 \, d\theta = 0$$

$$\int_{\Gamma_2} \mathbf{F} = \int_{2\pi}^0 \left(\frac{d}{dr}(r, 0, 2\pi) \right) \cdot (2\pi r, 0, -r^2) \, dr$$

$$= \int_{2\pi}^0 2\pi r \, dr = \frac{2\pi r^2}{2} \Big|_{2\pi}^0 = -\pi(2\pi)^2 = -4\pi^3$$

$$\int_{\Gamma_3} \mathbf{F} = \int_{2\pi}^0 \left(\frac{d}{d\theta}(0,0,\theta) \right) \cdot (0,0,0) \, d\theta = 0 \,.$$

Hence

$$\int_\Gamma \mathbf{F} = -4\pi^3 = \iint_S \mathrm{curl}(\mathbf{F})$$

and we have verified Stokes' theorem.

Moral: It is possible to verify Stokes' theorem.

EXERCISES

13.1 Let **S** denote the portion of the sphere $x^2 + y^2 + z^2 = 4a^2$ in the first octant which lies inside the cylinder $x^2 + y^2 = 2ax$ oriented outwards. Let Γ denote the boundary or edge of **S** directed in accordance with Stokes' theorem. Sketch **S** and Γ. Using Stokes' theorem evaluate

(a) $\displaystyle\int_\Gamma z\,dx - x\,dz,$

(b) $\displaystyle\int_\Gamma x\,dy - y\,dx,$

(c) $\displaystyle\int_\Gamma y\,dz - z\,dy.$

13.2 Sketch the surfaces $az = xy$ and $x^2 + y^2 = b^2$ in $\mathbb{R}^3$. Show that

$$\theta \longrightarrow (b\cos\theta, b\sin\theta, \frac{b^2\sin 2\theta}{2a}), \quad 0 \leq \theta \leq 2\pi$$

is a parametrization of the intersection Γ of the two surfaces oriented clockwise when viewed from a high point on the z-axis. Show that the surface parametrised by

$$P\colon (r, \theta) \longrightarrow (r\cos\theta, r\sin\theta, \frac{r^2\sin 2\theta}{2b}), \quad 0 < r < a,\ 0 < \theta < 2\pi$$

has Γ as its edge or boundary. Using Stokes' theorem find

$$\int_\Gamma y\,dx + z\,dy + x\,dz \,.$$

13.3 Let Γ denote the curve of intersection of $x + y = 2b$ and $x^2 + y^2 + z^2 = 2b(x + y)$ oriented in a clockwise sense when viewed from the origin. Sketch the appropriate diagram. Using Stokes' theorem evaluate

$$\int_\Gamma y\,dx + z\,dy + x\,dz \,.$$

13.4 Let S denote the torus obtained by rotating the circle $(x - b)^2 + z^2 = a^2$ in the xz-plane about the z-axis. Sketch S. Let Γ denote the directed curve parametrised by

$$t \longrightarrow \big((b + a\cos nt)\cos t, (b + a\cos nt)\sin t, a\sin nt\big) \,.$$

Show that Γ is a closed curve in S. Describe and sketch S. Let **R** denote the surface parametrised and oriented by

$$P\colon (r, t) \longrightarrow \big(r(b + a\cos nt)\cos t, r(b + a\cos nt)\sin t, a\sin nt\big)$$

where $0 \le r < 1$, $0 < t < 2\pi$. Show that Γ is the boundary of $\mathbf{R}$. Let $\mathbf{F}(x, y, z) = (-y, x, 0)$. By computing both $\int_\Gamma \mathbf{F}$ and $\int_\mathbf{R} \operatorname{curl}(\mathbf{F})$ verify Stokes' theorem. Show that the area of the projection of $\mathbf{R}$ onto $\mathbb{R}^2_{(x,y)}$ is $\pi(a^2 + 2b^2)/2$.

13.5 Use Stokes' theorem to evaluate

$$\iint_S \operatorname{curl}(y^2, xy, xz)$$

where S is the hemisphere $x^2 + y^2 + z^2 = 1$, $z > 0$, oriented so that the unit normal has positive z coordinate.

13.6 Let Γ denote the curve of intersection of the cylinder $x^2 + y^2 = a^2$ and the plane $x/a + z/b = 1$, $a > 0$, $b > 0$. Use Stokes' theorem to find a direction along Γ so that $\int_\Gamma (y - z, z - x, x - y)$ is positive. Find the value of this positive number.

13.7 Use Stokes' theorem to find a suitable orientation of the curve of intersection Γ of the hemisphere $x^2 + y^2 + z^2 = 2ax$, $z > 0$, and the cylinder $x^2 + y^2 = 2bx$, $0 < b < a$, so that

$$\int_\Gamma (y^2 + z^2)dx + (x^2 + z^2)dy + (x^2 + y^2)dz = 2\pi ab^2 .$$

13.8 Let $\phi = \tan^{-1}(\sqrt{x^2 + y^2}/z)$ and $\theta = \tan^{-1}(y/x)$ be latitude and longitude respectively on the unit sphere $x^2 + y^2 + z^2 = 1$. Let Γ_1 and Γ_2 denote the curves $\theta = 2\phi$ and $\theta = \phi + \pi$, $0 \le \phi \le \pi$, joining the North and South poles and let S denote the surface of the sphere between the two curves.

If $\mathbf{F} = (-yz\mathbf{i} + xz\mathbf{j})/(x^2 + y^2)(x^2 + y^2 + z^2)^{1/2}$ and $\mathbf{G} = \operatorname{curl}(\mathbf{F})$ find

$$\iint_S \langle F, G \rangle dA.$$

Verify your answer using Stokes' theorem.

14
Triple Integrals

Summary. *We define triple integrals of scalar-valued functions over open subsets of* $\mathbb{R}^3$, *discuss coordinate systems in* $\mathbb{R}^3$, *justify a change of variable formula and use Fubini's theorem to evaluate integrals.*

Let f be a real-valued function defined on an open subset U of $\mathbb{R}^3$. By using partitions of the coordinate axes to draw planes parallel to the coordinate planes (Figure 14.1) we obtain a *grid* which partitions $\mathbb{R}^3$ into cubes. Let $(\overline{x}_i, \overline{y}_j, \overline{z}_k)$

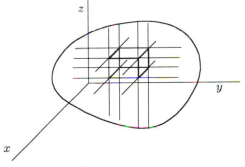

Figure 14.1.

denote a typical point in the cube $[x_i, x_{i+1}] \times [y_j, y_{j+1}] \times [z_k, z_{k+1}]$. The Riemann sum of f with respect to this grid

$$\sum_i \sum_j \sum_k f(\overline{x}_i, \overline{y}_j, \overline{z}_k)(x_{i+1} - x_i)(y_{j+1} - y_j)(z_{k+1} - z_k)$$

165

is formed by summing over all cubes that lie in U. If the Riemann sums converge, as we take finer and finer partitions and grids, to a limit then f is said to be *Riemann integrable* and the limit

$$\iiint_U f(x,y,z)dxdydz$$

is called the (Riemann) integral of f over U.

If f is a bounded open subset of $\mathbb{R}^3$ with smooth boundary and f is the restriction to U of a continuous function $\overline{f}$ on $\overline{U}$ then f is integrable over U. This result, proved using *uniform continuity* of $\overline{f}$ on the *compact* subset $\overline{U}$ of $\mathbb{R}^3$, implies the existence of an abundance of integrable functions.

If $f(x,y,z) \equiv 1$ the Riemann sum is the volume of all cubes inside U and in the limit equals the volume of U, Vol (U). Thus

$$\mathrm{Vol}(U) = \iiint_U dxdydz \,.$$

In evaluating triple integrals we use an extension of *Fubini's theorem*. This is obtained from the Riemann sum by first summing over i, taking a limit, then summing over j and taking a limit and finally summing over k and taking a limit. To justify this process it is usual to assume that the domain of integration has a "box-like" appearance, i.e. it is bounded above and below by surfaces $z = f_1(x,y)$ and $z = f_2(x,y)$, front and rear by surfaces $x = g_1(y,z)$ and $x = g_2(y,z)$ and on the left and right by surfaces $y = h_1(x,z)$ and $y = h_2(x,z)$. The situations we discuss are of this type but by no means reflect the full range of examples to which Fubini's theorem applies. Many open sets can be partitioned into a finite union of sets and Fubini's theorem applies to each of these. We refer to our remarks on Green's theorem in Chapter 9 for further details.

To apply Fubini's theorem we must examine various cross-sections of the domain of integration U. First suppose that the set of *non-empty cross-sections* of U parallel to the xy-plane, i.e. those obtained by fixing the z-coordinate, determine an interval (a,b) on the z-axis (Figure 14.2).

Let $A(z)$ denote the cross-section defined by fixing z in (a,b), i.e.

$$A(z) = \{(x,y) \in \mathbb{R}^2_{(x,y)} : (x,y,z) \in U\}\,.$$

This means

$$\iiint_U f(x,y,z)dxdydz = \int_a^b \Big\{ \iint_{A(z)} f(x,y,z)dxdy \Big\}dz$$

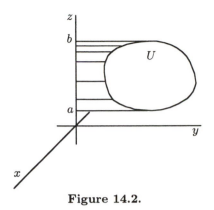

Figure 14.2.

and we now have to evaluate the inner integral of *two* variables. In some cases it is possible to do this directly, see for instance Example 14.1, but usually we apply the two-variable Fubini's theorem to each $A(z)$. For this we assume that the region $A(z) \subset \mathbb{R}^2_{(x,y)}$ is bounded on the left and right by the graphs of functions of y over some interval (Figure 14.3). Both the functions and the interval will depend on z and we denote the interval by $\big(c(z), d(z)\big)$ and the functions on the left and right by $y \to l(y, z)$ and $y \to r(y, z)$ respectively.

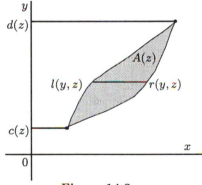

Figure 14.3.

This implies

$$\iint\limits_{A(z)} f(x, y, z) dx dy = \int_{c(z)}^{d(z)} \left\{ \int_{l(y,z)}^{r(y,z)} f(x, y, z) dx \right\} dy$$

and

$$\iiint_U f(x,y,z)dxdydz = \int_a^b \Big\{ \int_{c(z)}^{d(z)} \Big\{ \int_{l(y,z)}^{r(y,z)} f(x,y,z)dx \Big\} dy \Big\} dz \, .$$

Thus to evaluate triple integrals it is necessary to identify, by sketching, cross-sections of the open set U. Once this has been achieved and the result compared with the abstract figures above it is a matter of writing down the iterated integrals and evaluating them using one-variable integration theory. We have, as in the two-dimensional case a choice in the order of integration – in fact a total of $3! = 6$ choices, some may be easy, others difficult and some impossible. There are no definite rules.

A particularly simple situation occurs when $U = (a_1, b_1) \times (a_2, b_2) \times (a_3, b_3)$ and $f(x,y,z) = g(x)h(y)k(z)$ where g, h and k are functions of a single variable. In this case

$$\iiint_U f(x,y,z)dxdydz = \Big(\int_{a_1}^{b_1} f(x)dx \Big) \Big(\int_{a_2}^{b_2} g(y)dy \Big) \Big(\int_{a_3}^{b_3} k(z)dz \Big) \, .$$

For example, if $U = (0,a) \times (0,b) \times (0,c)$ then

$$\iiint_U xy^2 z^3 dxdydz = \int_0^a xdx \cdot \int_0^b y^2 dy \cdot \int_0^c z^3 dz = \frac{a^2 b^3 c^4}{24} \, .$$

Example 14.1

Let B denote the solid ball of radius r centred at the origin, i.e. $B = \{(x,y,z) : x^2 + y^2 + z^2 < r^2\}$. We calculate the volume of B by integrating the function $f \equiv 1$ over B. Figure 14.4(a) shows clearly that the values of z which give non-zero cross-sections lie in the interval $(-r, r)$ and the cross-section for fixed z is the disc $x^2 + y^2 \le r^2 - z^2$ (Figure 14.4(b)).

In this special case a direct computation is possible since

$$\iint_{A(z)} 1 dxdy = \text{Area}\big(A(z)\big) = \pi(r^2 - z^2)$$

and

$$\text{Vol}(B) \;=\; \int_{-r}^r \Big\{ \int_{A(z)} 1 dxdy \Big\} dz \tag{14.1}$$

$$=\; \int_{-r}^r \pi(r^2 - z^2)dz = \pi\Big(r^2 z - \frac{z^3}{3}\Big) \Big|_{-r}^r = \frac{4}{3}\pi r^3 \, . \tag{14.2}$$

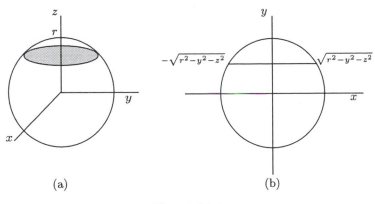

Figure 14.4.

We now consider the more typical approach to evaluating the inner integral over $A(z)$ by applying Fubini's theorem in two variables. In Figure 14.4(b) we have sketched the cross-section $A(z)$ in the xy-plane. The equation $x^2 + y^2 = r^2 - z^2$ (z fixed) has two solutions

$$x = \pm\sqrt{r^2 - z^2 - y^2}\,.$$

These give the total variation of x and the boundary functions, l and r. We thus have

$$
\begin{aligned}
\text{Vol}(B) &= \int_{-r}^{r}\left\{\int_{-\sqrt{r^2-z^2}}^{+\sqrt{r^2-z^2}}\left\{\int_{-\sqrt{r^2-y^2-z^2}}^{+\sqrt{r^2-y^2-z^2}} dx\right\}dy\right\}dz \\
&= \int_{-r}^{r}\left\{\int_{-\sqrt{r^2-z^2}}^{+\sqrt{r^2-z^2}} 2\sqrt{r^2-y^2-z^2}dy\right\}dz \\
&= 2\int_{-r}^{r}\left\{\int_{-\pi/2}^{+\pi/2}(r^2-z^2)\cos^2\theta d\theta\right\}dz \\
&= \pi\int_{-r}^{r}(r^2-z^2)dz = \pi\left(r^2 z - \frac{z^3}{3}\right)\Big|_{-r}^{r} = \frac{4\pi r^3}{3}
\end{aligned}
$$

where we let $y = (r^2 - z^2)^{1/2}\sin\theta$, $dy = (r^2 - z^2)^{1/2}\cos\theta$.

The geometry of the previous example was reasonably straightforward. Many examples appear initially to involve a rather complicated geometric shape. However, we are usually dealing with a limited number of objects, mainly conic sections, mixed together and once familiarity with these has been established and sufficiently many cross-sections sketched – with the help of the defining inequalities – the correct approach often presents itself.

Example 14.2

We wish to find the volume of the region V lying below the plane $z = 3-2y$ and above the paraboloid $z = x^2 + y^2$, i.e. the set $\{(x, y, z) : x^2 + y^2 < z < 3 - 2y\}$. We begin by considering a full sketch (Figure 14.5).

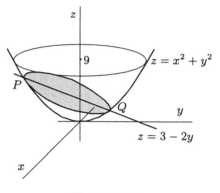

Figure 14.5.

The coordinates of P and Q are found by solving between the equations $z = x^2 + y^2$ and $z = 3 - 2y$. These imply $x^2 + y^2 = 3 - 2y$, i.e. $x^2 + y^2 + 2y + 1 = 4$. Hence $x^2 + (y+1)^2 = 2^2$. The extreme values of y are obtained by letting $x = 0$. This gives $(y+1)^2 = 2^2$, i.e. $y + 1 = \pm 2$. Hence $y = -3$ or $+1$. The coordinates of P and Q are $(0, -3, 9)$ and $(0, 1, 1)$ respectively. Hence $-3 \leq y \leq +1$. From Figure 14.5 we see $0 \leq z \leq 9$ and from the equation $x^2 + (y+1)^2 = 2$ we deduce that $-2 \leq x \leq 2$. Having found the extremal values for non-empty sections we sketch the corresponding cross-sections.

First fix y, $-3 \leq y \leq 1$. In the zx-plane, $z = y^2 + x^2$ is a parabola and $z = 3 - 2y$ is a straight line. From Figure 14.6 it follows that for fixed y,

$$-(3 - 2y - y^2)^{1/2} \leq x \leq (3 - 2y - y^2)^{1/2}$$

and for fixed x and y

$$x^2 + y^2 \leq z \leq 3 - 2y.$$

We have

$$\mathrm{Vol}(V) \;=\; \int_{-3}^{1} \left\{ \int_{-(3-2y-y^2)^{1/2}}^{(3-2y-y^2)^{1/2}} \left\{ \int_{x^2+y^2}^{3-2y} dz \right\} dx \right\} dy$$

$$=\; \int_{-3}^{1} \left\{ \int_{-(3-2y-y^2)^{1/2}}^{(3-2y-y^2)^{1/2}} [z]_{x^2+y^2}^{3-2y}\, dx \right\} dy$$

$$= \int_{-3}^{1} \left\{ \int_{-(3-2y-y^2)^{1/2}}^{(3-2y-y^2)^{1/2}} (3 - 2y - y^2 - x^2)dx \right\} dy$$

$$= \int_{-3}^{1} \left[(3 - 2y - y^2)x - \frac{x^3}{3} \right]_{-(3-2y-y^2)^{1/2}}^{(3-2y-y^2)^{1/2}} dy$$

$$= \int_{-3}^{1} \frac{4}{3}(3 - 2y - y^2)^{3/2} dy = \frac{4}{3} \int_{-3}^{1} \left(4 - (y + 1)^2 \right)^{3/2} dy$$

Let $y + 1 = 2 \sin \theta$, then $dy = 2 \cos \theta d\theta$, $4 - (y+1)^2 = 4 - 4 \sin^2 \theta = 4 \cos^2 \theta$, so

$$\begin{aligned}
\text{Vol}(V) &= \frac{4}{3} \int_{-\pi/2}^{\pi/2} (4 \cos^2 \theta)^{3/2} \cdot 2 \cos \theta d\theta \\
&= \frac{64}{3} \int_{-\pi/2}^{\pi/2} \cos^4 \theta d\theta \\
&= \frac{64}{3} \int_{-\pi/2}^{\pi/2} \frac{1}{4} \left(1 + 2 \cos 2\theta + \frac{1}{2}(1 + \cos 4\theta) \right) d\theta \\
&= \frac{16}{3} \cdot \frac{3\pi}{2} = 8\pi .
\end{aligned}$$

We choose to take y as the final variable in our order of integration since it is clear from Figure 14.5 that all cross-sections parallel to the xz-plane are of the same type whereas cross-sections parallel to the xy-plane, i.e. fixing z, are different for $z < 1$ and $z \geq 1$. A different order of integration can be used as a second opinion.

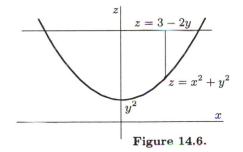

Figure 14.6.

From the two previous examples we see that the evaluation of triple integrals proceeded in three stages. First we chose an order of integration and next examined the geometry of the domain of integration in order to determine the limits of integration in the inner integrals. Finally we evaluated a sequence of one variable integrals. The alternatives at each stage in the process point towards a useful technique commonly called *change of variable*. In place of a detailed motivation we briefly mention three pertinent ideas.

(a) The domain of integration U was partitioned into cubes, Figure 14.7(a), with element of volume ΔV easily calculated using $\Delta x \times \Delta y \times \Delta z$. We could use instead a grid based on spheres centred at the origin and planes through the origin to obtain a different element of volume, Figure 14.7(b), or a grid based on cylinders parallel to the z-axis and planes perpendicular and parallel to the z-axis (Figure 14.7(c)).

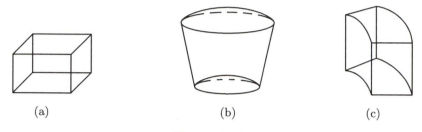

(a)   (b)   (c)

Figure 14.7.

These alternatives lead to more complicated formulae for ΔV but hopefully the new limits of integration and the resulting one-variable integrals are less complicated.

(b) We used Cartesian coordinates (x, y, z) to denote a typical point in the domain of integration U but we may consider other methods of identifying points in U. For instance, if U is the solid sphere of radius 1 then each point in U lies in a sphere of radius r, $0 \le r \le 1$, and using the parametrization of the sphere of radius r given in Example 10.3 we can identify points of U by means of r (the distance to the origin), θ the angle of latitude and ψ (the angle of longitude). In terms of the coordinates (r, θ, ψ) the domain U becomes the parallelepiped $(0, 1) \times (-\pi/2, \pi/2) \times (0, 2\pi)$ and, as previously noted, integration in this case is much more pleasant. The new set of coordinates gives a correspondence F between a domain in $\mathbb{R}^3_{(r,\theta,\psi)}$ and the original domain U in $\mathbb{R}^3_{(x,y,z)}$ (Figure 14.8).

By Example 10.3, $F(r, \theta, \psi) = (r \cos \theta \cos \psi, r \cos \theta \sin \psi, r \sin \theta)$. The idea now is to transfer the cubical grid on U, using F, to a grid, which is usually not cubical, on $F(U)$ and hence to evaluate the integral. In carrying out this operation it will be necessary to calculate

$$\mathrm{Vol}\big(F(\Delta V)\big) = \mathrm{Vol}\big(F(\Delta r \times \Delta \theta \times \Delta \psi)\big)$$

and this is the problem that also arises in (a). The mapping F has many of the features that we have previously associated with a parametrization and, by now, the following definition should appear natural.

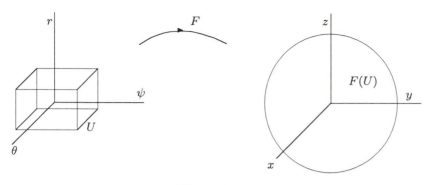

Figure 14.8.

Definition 14.1

A parametrization of an open set V in $\mathbb{R}^n$ is a bijective differentiable mapping F from an open subset U of $\mathbb{R}^n$ onto V such that $F'(X)$ is an invertible linear operator for all X in U.

This definition contains the essential properties that we used to parametrise curves and surfaces and may also be regarded as a method of providing V with a *new coordinate system*. The requirement that $F'(X)$ be invertible (or equivalently that $\det\big(F'(X)\big) \neq 0$) is the three-dimensional analogue of the condition $P'(t) \neq 0$ for curves and $\phi_x \times \phi_y \neq 0$ for surfaces.

(c) A third, perhaps more obscure, approach is motivated by the substitutions that may arise in the one-variable integrals in the final stage. In Example 14.2 we needed one such change. Working backwards it may be possible to choose initially a coordinate system which does not require a change of variable in the iterated integrals.

We return now to (b) above to work out the formula for the change of variable. Let U denote an open subset of $\mathbb{R}^3_{(r,s,t)}$ and let $g\colon U \to g(U) = V$ denote a parametrization of the open subset V of $\mathbb{R}^3_{(u,v,w)}$. (Note that g^{-1} is a parametrization of U.) To avoid confusion we think of U and V as lying in different copies of $\mathbb{R}^3$ each with their own set of coordinates, (r,s,t) and (u,v,w) respectively. This explains the terminology "change of variables".

Let f denote an integrable function on $g(U)$ (Figure 14.9).

Take a cubical grid on U, transfer it by g to a grid on $g(U)$ and then form a Riemann sum of f. A typical term in this Riemann sum is

$$f\big(g(\overline{r}_i, \overline{s}_j, \overline{t}_k)\big)\,\mathrm{Vol}\big(g(\Delta r_i \times \Delta s_j \times \Delta t_k)\big)\,.$$

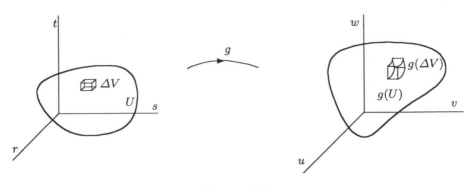

Figure 14.9.

Now $g(\Delta r_i \times \Delta s_j \times \Delta t_k)$ is approximately a parallelepiped with sides $g_r\Delta r$, $g_s\Delta s$ and $g_t\Delta t$. We have already discussed the volumes of parallelepipeds while introducing Stokes' theorem in Chapter 12. The area of the base of the parallelepiped is $\|g_r \times g_s\Delta r_i \cdot \Delta s_j\|$ (Figure 14.10(a)) where the partial derivatives of g are evaluated at $(\bar{r}_i, \bar{s}_j, \bar{t}_k)$. The height of the parallelepiped is the length of the projection of $g_t\Delta t_k$ onto a direction perpendicular to the base. Since $g_r \times g_s$ is perpendicular to the base the required height is $\left(g_r \times g_s \cdot g_t \, / \, \|g_r \times g_s\|\right)\Delta t_k = \|g_t\| \cos\theta \Delta t_k$ (Figure 14.10(b)). Hence

$$\mathrm{Vol}\big(g(\Delta r_i \times \Delta s_j \times \Delta t_k)\big) \;\approx\; \|g_r \times g_s\Delta r_i\Delta s_j\| \cdot \left| g_t \frac{g_r \times g_s}{\|g_r \times g_s\|}\right| \Delta t_k$$

$$= \;\|g_r \times g_s \cdot g_t\|\Delta r_i\Delta s_j\Delta t_k \, .$$

An easy exercise shows that

$$\|g_r \times g_s \cdot g_t\| = \left| \det \begin{pmatrix} g_r \\ g_s \\ g_t \end{pmatrix}\right| = \big|\det(g')\big|$$

and

$$\mathrm{Vol}\,\big(g(\Delta r_i \times \Delta s_j \times \Delta t_k)\big) \approx \big|\det(g')\big|\Delta r_i\Delta s_j\Delta t_k \, .$$

Hence the Riemann sum of f over $g(U)$ is approximately

$$\sum_i \sum_j \sum_k f\big(g(\bar{r}_i, \bar{s}_j, \bar{t}_k)\big)\big|\det\big(g'(\bar{r}_i, \bar{s}_j, \bar{t}_k)\big)\big|\Delta r_i\Delta s_j\Delta t_k$$

where we sum over the cubes in the partition of U. In the limit we get the *change of variables formula*

$$\iiint\limits_{g(U)} f(u, v, w)\,du\,dv\,dw = \iiint\limits_{U} f\big(g(r, s, t)\big)\big|\det\big(g'(r, s, t)\big)\big|\,dr\,ds\,dt \, .$$

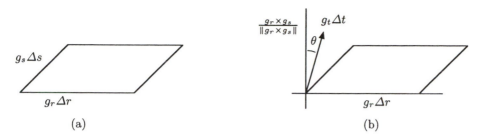

Figure 14.10.

The notations $J(g)$ and $\dfrac{\partial(u, v, w)}{\partial(r, s, t)}$ are also used in place of $\det(g')$ and this determinant is called the *Jacobian* of g.

If V is an open subset of $\mathbb{R}^2$ then we may identify V with the open subset $\widetilde{V} = V \times (0, 1)$ in $\mathbb{R}^3$. A function $f: V \to \mathbb{R}$ is integrable over V if and only if $\widetilde{f}$, defined by $\widetilde{f}(x, y, z) = f(x, y)$, is integrable over $\widetilde{V}$ and, moreover,

$$\iiint_{\widetilde{V}} \widetilde{f}(x, y, z)\, dx dy dz = \iint_V f(x, y)\, dx dy\,.$$

If $g: U \subset \mathbb{R}^2_{(r,s)} \to V \subset \mathbb{R}^2_{(u,v)}$ is a mapping between the open sets U and V then it is easily seen that g is a parametrization of V if and only if $\widetilde{g}: \widetilde{U} \subset \mathbb{R}_{(r,s,t)} \to \widetilde{V} \subset \mathbb{R}_{(u,v,w)}$, defined by $\widetilde{g}(r, s, t) = \big(g(r, s), t\big)$, is a parametrization of $\widetilde{V}$. Since

$$\widetilde{g}' = \begin{pmatrix} & & 0 \\ & g' & 0 \\ 0 & 0 & 1 \end{pmatrix}$$

we have $\det(g') = \det(\widetilde{g}')$ and

$$
\begin{aligned}
\iint_V f \;&=\; \iiint_{\widetilde{V}} \widetilde{f} = \iiint_{\widetilde{U}} \widetilde{f} \circ \widetilde{g} |\det(\widetilde{g}')| \\
&=\; \iint_U f \circ g |\det(g')|\,.
\end{aligned}
$$

This justifies the change of variables formula for double integrals (Chapter 9) and yields the following familiar formula

$$\iint_{V = g(U)} f(u, v)\, du dv = \iint_U f\big(g(r, s)\big) \begin{vmatrix} u_r & u_s \\ v_r & v_s \end{vmatrix} dr ds$$

where $g(r,s) = (u(r,s), v(r,s))$. In particular, for polar coordinates in the plane, $g: (r, \theta) \to (x, y) = (r \cos \theta, r \sin \theta)$, we have

$$|\det(g')| = \begin{vmatrix} \cos \theta & \sin \theta \\ -r \sin \theta & r \cos \theta \end{vmatrix} = r$$

and

$$\iint\limits_{x^2 + y^2 < 1} f(x, y)\,dxdy = \iint\limits_{\substack{0 < r < 1 \\ 0 < \theta < 2\pi}} f(r \cos \theta, r \sin \theta) r\, dr d\theta \,.$$

If **S** is a simple oriented surface in $\mathbb{R}^3$ and (g_1, U_1) and (g_2, U_2) are parametrizations of S consistent with the orientation then the bijectivity and smoothness of g_1 and g_2 imply that the mapping $g_2^{-1} \circ g_1$ is a parametrization of U_2 (Figure 14.11).

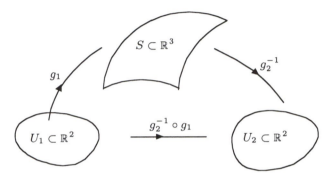

Figure 14.11.

The change of variables formula for double integrals may now be used to show that integrals over **S** are *independent* of the parametrization and justifies the notation used in earlier chapters (we remark that g_1 and g_2 map onto the *same* side of **S** if and only if $\det\big((g_2^{-1} \circ g_1)'\big)$ is always *strictly positive*).

We have noted above how a solid sphere is a union of surfaces, i.e.

$$\{(x, y, z) : x^2 + y^2 + z^2 < r^2\} = \bigcup_{0 \le s < r} \{(x, y, z) : x^2 + y^2 + z^2 = s^2\}$$

and using parametrization of surfaces we were able to fill in to obtain a parametrization of the solid sphere. In fact, we do not quite get a parametrization of the full solid, since to obtain a bijective mapping, we are forced to miss a small portion of the sphere. This is the same problem that we encountered and discussed fully in parametrizing the classical surfaces (Chapter 10). We do not enter into a full discussion here but remark that after a similar discussion we would arrive at an analogous conclusion for solids. The portion of the

solid omitted has volume zero and so for integration purposes we may treat the
mappings in the following example as parametrizations of the full solid.

Example 14.3

Parametrization of solids in $\mathbb{R}^3$.

(a) *Sphere of radius a* (spherical polar coordinates)

$$(0, a) \times (0, \pi) \times (0, 2\pi) \in \mathbb{R}^3 \longrightarrow \left\{(x, y, z) : x^2 + y^2 + z^2 < a^2\right\}$$

$$(r, \theta, \psi) \longrightarrow (r \sin \theta \cos \psi, r \sin \theta \sin \psi, r \cos \theta)$$

(b) *Ellipsoid* (elliptical polar coordinates)

$$(0, 1) \times (0, \pi) \times (0, 2\pi) \in \mathbb{R}^3 \longrightarrow \left\{(x, y, z) : \frac{x^2}{a^2} + \frac{y^2}{b^2} + \frac{z^2}{c^2} < 1\right\}$$

$$(r, \theta, \psi) \longrightarrow (ra \sin \theta \cos \psi, rb \sin \theta \sin \psi, c \cos \theta)$$

(c) *Solid of revolution* generated by revolving the area beneath the plane curve
$P(t) = (x(t), y(t)), t \in [a, b]$ and $y(t) > 0$, about the x–axis (Example 10.2),

$$(0, 1) \times (a, b) \times (0, 2\pi) \in \mathbb{R}^3 \longrightarrow \text{Solid of Revolution}$$

$$(r, t, \theta) \longrightarrow (x(t), ry(t) \cos \theta, ry(t) \sin \theta)$$

(d) The *cylinder* of radius r and height h parallel to the z-axis is defined by
$\{(x, y, z) : x^2 + y^2 < r^2, \, 0 < z < h\}$ and parametrised by

$$(s, \theta, z) \longrightarrow (s \cos \theta, s \sin \theta, z)$$

with domain $(0, r) \times (0, 2\pi) \times (0, h)$ (cylindrical coordinates)

(e) The inverted solid *cone* $\{(x, y, z) : x^2 + y^2 < z^2, \, 0 < z < 1\}$ is parametrised
by

$$(r, \theta, z) \longrightarrow (r \cos \theta, r \sin \theta, z)$$

where $0 < r < z$, $0 < \theta < 2\pi$, $0 < z < 1$.

(f) The inverted solid *paraboloid* defined by $\{(x, y, z) : x^2 + y^2 < z, \, 0 < z < 1\}$
is parametrised by

$$(r, \theta, z) \longrightarrow (r \cos \theta, r \sin \theta, z)$$

where $0 < r < \sqrt{z}$, $0 < \theta < 2\pi$, $0 < z < 1$.

Example 14.4

In this example we calculate the volume of a solid *torus*. We discussed the boundary surface of a torus in Example 11.1 and obtained the parametrization

$$f(\theta, \psi) = \Big((b + r\cos\theta)\cos\psi, (b + r\cos\theta)\sin\psi, r\sin\theta\Big)$$

where $0 < \theta < 2\pi$, $0 < \psi < 2\pi$ (Figure 11.4). We generate the *solid torus* by rotating the disc and obtain the parametrization

$$F\colon (0, 2\pi) \times (0, 2\pi) \times (0, r) \longrightarrow \text{Solid Torus}$$

$$(\theta, \psi, s) \longrightarrow \Big((b + s\cos\theta)\cos\psi, (b + s\cos\theta)\sin\psi, s\sin\theta\Big).$$

We will use the change of variable formula to calculate the volume but first note how the following geometric observation immediately gives the answer. The solid torus is obtained by rotating a disc of radius r with centre on the y-axis at a distance b from the origin about the z-axis. Thus the disc is rotated through a distance $2\pi b$ and generates a solid whose volume is

$$\pi r^2 \cdot 2\pi b = 2\pi^2 r^2 b.$$

From the change of variable formula

$$\text{Volume (Torus)} = \iiint\limits_{[0,2\pi] \times [0,2\pi] \times [0,1]} \Big|\det\big(F'(\theta, \psi, s)\big)\Big| d\theta d\psi ds.$$

We have

$$F'(\theta, \psi, s) = \begin{pmatrix} -s\sin\theta\cos\psi & -(b + s\cos\theta)\sin\psi & \cos\theta\cos\psi \\ -s\sin\theta\sin\psi & (b + s\cos\theta)\cos\psi & \cos\theta\sin\psi \\ s\cos\theta & 0 & \sin\theta \end{pmatrix}.$$

From the matrix representation it is easily seen that F_θ, F_ψ and F_s, i.e. the columns of F', are three mutually perpendicular vectors and hence generate a parallelepiped shaped like a rectangular box. In this case the volume is the product of the lengths of the sides. Hence

$$
\begin{aligned}
\Big|\det\big(F'(\theta, \psi, s)\big)\Big| &= \|F_\theta\| \cdot \|F_\psi\| \cdot \|F_s\| \\
&= (s^2\sin^2\theta\cos^2\psi + s^2\sin^2\theta\sin^2\psi + s^2\cos^2\theta)^{1/2} \cdot \\
&\quad \big((b + s\cos\theta)^2\sin^2\psi + (b + s\cos\theta)^2\cos^2\psi\big)^{1/2} \cdot \\
&\quad (\cos^2\theta\cos^2\psi + \cos^2\theta\sin^2\psi + \sin^2\theta)^{1/2} \\
&= s(b + s\cos\theta)
\end{aligned}
$$

and

$$
\begin{aligned}
\text{Volume (Torus)} &= \int_{[0,2\pi]} \int_{[0,2\pi]} \int_0^r s(b + s\cos\theta)\,d\theta\,d\psi\,ds \\
&= \int_0^{2\pi} d\psi \cdot \int_0^r \left\{ \int_0^{2\pi} (sb + s^2\cos\theta)\,d\theta \right\} ds \\
&= 2\pi \int_0^r (sb\theta + s^2\sin\theta) \Big|_0^{2\pi} ds \\
&= 2\pi \int_0^r sb2\pi\,ds = 4\pi^2 b \frac{s^2}{2} \Big|_0^r = 2\pi^2 br^2\,.
\end{aligned}
$$

Example 14.5

To find the volume of the solid contained within the sphere $x^2 + y^2 + z^2 = 4a^2$ and the cylinder $x^2 + (y-a)^2 = a^2$. In Example 12.3 we considered a geometric situation similar to the present one. By not moving the origin we have adopted a slightly different approach here. From Figure 14.12 we see that the solid lies above and below the plane disc $x^2 + (y-a)^2 \leq a^2$ and this immediately suggests polar coordinates.

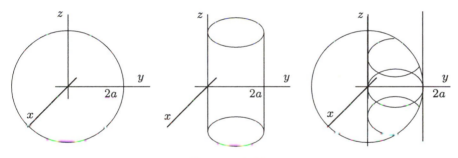

Figure 14.12.

The volume V is equal to

$$
\iiint_{\substack{x^2+y^2+z^2<4a^2 \\ x^2+(y-a)^2<a^2}} 1\,dx\,dy\,dz\,.
$$

We parametrise the solid using *polar coordinates* in the xy-plane and the usual Cartesian z coordinate, i.e. we use the *cylindrical coordinates*

$$
F\colon (r,\theta,z) \longrightarrow (r\cos\theta, r\sin\theta, z)\,.
$$

Since

$$F'(r,\theta,z) = \begin{pmatrix} \cos\theta & -r\sin\theta & 0 \\ \sin\theta & r\cos\theta & 0 \\ 0 & 0 & 1 \end{pmatrix}$$

we have $\det\big(F'(r,\theta,z)\big) = r$. To find the limits of integration consider Figure 14.13 in the xy-plane.

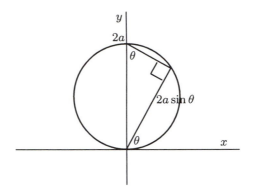

Figure 14.13.

We see that $0 < \theta < \pi$, $0 < r < 2a\sin\theta$ and, from Figure 14.12, $z^2 < 4a^2 - x^2 - y^2 = 4a^2 - r^2$, i.e. $-\sqrt{4a^2 - r^2} < z < \sqrt{4a^2 - r^2}$. Hence

$$\text{Volume} = \int_0^\pi \Big\{ \int_0^{2a\sin\theta} \Big\{ \int_{-\sqrt{4a^2-r^2}}^{\sqrt{4a^2-r^2}} r\,dz \Big\} dr \Big\} d\theta$$

$$= 2\int_0^\pi \Big\{ \int_0^{2a\sin\theta} r\sqrt{4a^2 - r^2}\,dr \Big\} d\theta.$$

Let $s = 4a^2 - r^2$. Then $ds = -2r\,dr$ and

$$\text{Volume} = 2\int_0^\pi \Big\{ \int_{4a^2}^{4a^2\cos^2\theta} \big(-\tfrac{1}{2}s^{1/2}\big)ds \Big\} d\theta$$

$$= -\int_0^\pi \Big(\frac{2s^{3/2}}{3} \Big|_{4a^2}^{4a^2\cos^2\theta} \Big) d\theta$$

$$= \frac{16a^3}{3}\int_0^\pi (1 - \cos^3\theta)\,d\theta = \frac{16a^3\pi}{3}.$$

Example 14.6

In this example we calculate the volume of the region U bounded above by the paraboloid $z = 9 - x^2 - y^2$, below by the xy-plane and which lies outside the cylinder $x^2 + y^2 = 1$ (Figure 14.14).

The presence of $x^2 + y^2$ in the defining inequalities suggests cylindrical coordinates (r, θ, z) (the presence of $x^2 + y^2 + z^2$ would suggest geographical or spherical polar coordinates). From Figure 14.14 we see that U projects onto $\{(x, y) : 1 \leq x^2 + y^2 \leq 9\}$ in the xy-plane. Hence $1 \leq r \leq 3$ and z varies over $0 \leq z \leq 9 - x^2 - y^2 = 9 - r^2$. From the previous example we know that the Jacobian is equal to r. Hence the required volume is

$$\int_0^{2\pi} \{\int_1^3 \{\int_0^{9-r^2} r\, dz\} dr\} d\theta \;=\; \int_0^{2\pi} \{\int_1^3 (9r - r^3) dr\} d\theta$$

$$=\; 2\pi \cdot \left(\frac{9r^2}{2} - \frac{r^4}{4}\right)\Big|_1^3 = 32\pi \,.$$

EXERCISES

14.1 Show that the volume of the solid inside the cylinder $x^2 + y^2 - 2ay = 0$ and between the plane $z = 0$ and the paraboloid $4az = x^2 + y^2$ equals $3\pi a^3/8$.

14.2 Find the volume of the solid inside the cylinder $x^2 + y^2 = 2ax$ which lies between the plane $z = 0$ and the cone $x^2 + y^2 = z^2$.

14.3 Show that the volume of the solid defined by the inequalities $x^2 + y^2 \leq 1$ and $\tan^{-1}(y/x) \leq z \leq 2\pi$ equals π^2.

14.4 Let U denote the region above the plane $z = 0$ between the cone $z^2 = x^2 + y^2$ and the paraboloid $z = 2 - x^2 - y^2$. Show that this region projects onto the unit disc in the xy-plane. Using cylindrical coordinates or otherwise show that the volume of U equals $5\pi/6$.

14.5 Evaluate the following integrals:

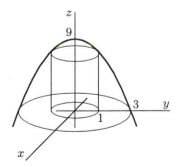

Figure 14.14.

(a) $\displaystyle\int_{-3}^{1}\{\int_{y^2}^{3-2y}\{\int_{-(z-y^2)^{1/2}}^{(z-y^2)^{1/2}}dx\}dz\}dy$

(b) $\displaystyle\int_{-2}^{2}\{\int_{-1-(4-x^2)^{1/2}}^{-1+(4-x^2)^{1/2}}\{\int_{x^2+y^2}^{3-2y}dz\}dy\}dx$

(c) $\displaystyle\int_{0}^{9}\{\int_{-\sqrt{z}}^{\sqrt{z}}\{\int_{-(z-x^2)^{1/2}}^{\frac{3-z}{2}}dy\}dx\}dz$.

14.6 Use the change of variables formula to calculate the volumes of the solids parametrised in Example 14.3.

14.7 Find the volume of the wedge of the cylinder $\{(x,y,z) : x^2+y^2 \le 1\}$ which lies above the xy-plane and between the planes $z = -y$ and $z = 0$.

14.8 Find the volume of the open set which lies above the square $0 < x < 1$, $0 < y < 1$ in the xy-plane and below the surface $z = (1+x+y)^{1/2}$. Hence write down the volume of the set enclosed by the surfaces $z = (1+|x|+|y|)^{1/2}$, $z = -(1+|x|+|y|)^{1/2}$ and by the planes $x = 1$, $x = -1$, $y = 1$ and $y = -1$.

14.9 The value of the following integral is the volume of a region in $\mathbb{R}^3$. Sketch the region.

$$\int_{0}^{a}\{\int_{0}^{\sqrt{a^2-x^2}}\{\int_{\sqrt{3(x^2+y^2)}}^{\sqrt{4a^2-x^2-y^2}}dz\}dy\}dx .$$

14.10 Evaluate

$$\iiint_{V} z\,dxdydz$$

where $V = \{(x,y,z) : 0 < z < x\sqrt{y},\ 0 < y < 1,\ 1 < x < 2\}$.

14.11 Use the change of variables

$$(u,v,w) \longrightarrow \big(u(1-v),\, uv(1-w),\, uvw\big)$$

to calculate

$$\iiint_{V} x\,dxdydz \quad \text{and} \quad \iiint_{V} \frac{dxdydz}{y+z}$$

where V is the tetrahedron cut from the first octant by the plane $x + y + z = 1$.

14.12 Let Ω denote the region bounded by the rays $\theta = a$ and $\theta = b$ and the curve $r = f(\theta)$. Show that

$$\text{Area}\,(\Omega) = \frac{1}{2}\int_a^b f^2(\theta)d\theta.$$

Sketch the curve $r = 1 + \cos\theta$ and show that it encloses a region with area $3\pi/2$.

15

The Divergence Theorem

Summary. *We state, discuss and give examples of the divergence theorem of Gauss.*

The *divergence theorem of Gauss* is an extension to $\mathbb{R}^3$ of the fundamental theorem of calculus and of Green's theorem and is a close relative, but not a direct descendent, of Stokes' theorem. This theorem allows us to evaluate the integral of a scalar-valued function over an open subset of $\mathbb{R}^3$ by calculating the surface integral of a certain vector field over its boundary.

In Chapter 6 we defined the *divergence* of the vector field $\mathbf{F} = (f_1, f_2, f_3)$ as

$$\text{div}(\mathbf{F}) = \nabla \cdot \mathbf{F} = \left(\frac{\partial}{\partial x}, \frac{\partial}{\partial y}, \frac{\partial}{\partial z}\right) \cdot (f_1, f_2, f_3) = \frac{\partial f_1}{\partial x} + \frac{\partial f_2}{\partial y} + \frac{\partial f_3}{\partial z}$$

and we have previously written symbolically

$$\text{curl}(\mathbf{F}) = \nabla \times \mathbf{F}.$$

Carrying this symbolism a step further we now write

$$\text{div (curl } \mathbf{F}) = \nabla \cdot \nabla \times \mathbf{F} = \begin{vmatrix} \dfrac{\partial}{\partial x} & \dfrac{\partial}{\partial y} & \dfrac{\partial}{\partial z} \\ \dfrac{\partial}{\partial x} & \dfrac{\partial}{\partial y} & \dfrac{\partial}{\partial z} \\ f_1 & f_2 & f_3 \end{vmatrix}$$

This suggests, since the determinant of a matrix with two identical rows is zero, that $\text{div(curl } \mathbf{F}) = 0$ for any smooth vector field $\mathbf{F}$. This is indeed true. Symbolism, however, does *not* prove anything and it is necessary to verify this formally. At the same time it is a good example of the role of symbolism (and notation) in mathematics – it can be suggestive – and sometimes leads to true results that might otherwise be overlooked.

If $\mathbf{F}$ is the velocity of a gas then div($\mathbf{F}$) represents the rate of expansion (or compression) per unit volume. The divergence theorem states that the total expansion (or contraction) over a region U equals the total inflow (or outflow) across the boundary. Very little physical intuition is required in order to accept this as reasonable. This physical interpretation is responsible for the terminology "divergence".

We now formally state the divergence theorem.

Theorem 15.1 (Gauss' Divergence Theorem)

Suppose $\mathbf{S}$ is an oriented surface in $\mathbb{R}^3$, with outward pointing normal, enclosing an open set U and $\mathbf{F}$ is a smooth vector field on $\overline{U}$ ($= U \cup S$) then

$$\iint_{\mathbf{S}} \mathbf{F} = \iiint_{U} \mathrm{div}\ (\mathbf{F}).$$

An initial proof uses Green's theorem and Fubini's theorem on cross-sections of U and, as noted in earlier chapters, this places certain geometrical restrictions on U and S. From the previous chapter we see that it suffices to have U bounded, top and bottom, above and below, front and back by graphs of functions which form the boundary S. This covers many examples involving classical Euclidean shapes.

Further examples, for instance the annulus $0 < r_1 < x^2 + y^2 + z^2 < r_2 < \infty$, are obtained by dividing the open set into a finite number of sets on each of which the initial proof applies (we refer to our discussion on Green's theorem in Chapter 9 for further details). In general, the boundary will be composed of a finite number of distinct simple surfaces.

The collection of sets to which the divergence theorem applies is quite large but difficult to formulate precisely without recourse to further concepts from differential geometry. For this reason we have carefully avoided proving the theorem and giving detailed hypotheses on U and S in the statement of the theorem. The theorem, as stated, is sufficient for the examples we consider.

Example 15.1

In this example we use the divergence theorem to evaluate

$$\iiint_V \text{div } (\mathbf{F}) dxdydz$$

where V is the solid cylinder $\{(x, y, z) : x^2 + y^2 < 1, 0 < z < 1\}$ and

$$\mathbf{F}(x, y, z) = \left(1 - (x^2 + y^2)^3, 1 - (x^2 + y^2)^3, x^2 z^2\right) .$$

The boundary of V is composed of the cylinder $\{(x, y, z) : x^2 + y^2 = 1, 0 < z < 1\}$ and the flat discs $\{(x, y, z) : x^2 + y^2 \le 1, z = 0\}$ and $\{(x, y, z) : x^2 + y^2 \le 1, z = 1\}$ (Figure 15.1).

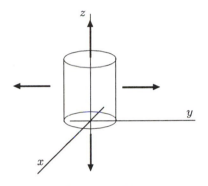

Figure 15.1.

The outward normal on the curved surface at the point (x, y, z) is $(x, y, 0)$. On the upper and lower discs the outward normals are $(0, 0, 1)$ and $(0, 0, -1)$ respectively. On the curved surface $\mathbf{F} \cdot \mathbf{n} = 0$ since $x^2 + y^2 = 1$ and the third coordinate of $\mathbf{n}$ is zero, on the bottom $\mathbf{F} \cdot \mathbf{n} = 0$ since $z = 0$ and on the top $\mathbf{F} \cdot \mathbf{n} = x^2$. By the divergence theorem

$$\iiint_V \text{div } (\mathbf{F}) dxdydz = \iint_{x^2+y^2 \le 1} x^2 dxdy .$$

The double integral is evaluated by using the polar coordinates

$$(r, \theta) \longrightarrow (r \cos \theta, r \sin \theta), \quad 0 < r < 1, 0 < \theta < 2\pi$$

on the unit disc and equals $\pi/4$.

Example 15.2

We use the divergence theorem to evaluate

$$\iint_{\partial \mathbf{W}} (x^2 + y + z)$$

where W is the solid ball $x^2 + y^2 + z^2 < 1$ and $\partial \mathbf{W}$ is its boundary oriented outwards. The unit sphere has outward normal $\mathbf{n}(x, y, z) = (x, y, z)$. If $\mathbf{F} = (f_1, f_2, f_3)$ is a vector field on $\overline{W}$ then the divergence theorem implies that

$$\iint_{\partial \mathbf{W}} (f_1, f_2, f_3) \quad = \quad \iint_{\partial W} (xf_1 + yf_2 + zf_3)dA \qquad (15.1)$$

$$= \quad \iiint_{W} (\frac{\partial f_1}{\partial x} + \frac{\partial f_2}{\partial y} + \frac{\partial f_3}{\partial z})dxdydz . \qquad (15.2)$$

Hence we can use any smooth vector field on the closed unit ball for which $xf_1 + yf_2 + zf_3 = x^2 + y + z$. This allows a wide choice but in such cases the simplest and most obvious option is usually the best. Let $f_1(x, y, z) = x$, $f_2(x, y, z) = 1$ and $f_3(x, y, z) = 1$. Then

$$\iint_{\partial W} (x^2 + y + z)dA = \iiint_{W} 1 dxdydz = \text{Vol } (W) = \frac{4}{3}\pi .$$

Example 15.3

In this example we verify a particular case of the divergence theorem. Let U denote the set defined by the inequalities $z \geq 0$, $x^2 + y^2 + z^2 \leq 4$ and $x^2 + y^2 \leq z^2$. These define the set of points, above the xy-plane, contained in the sphere of radius 2 centred at the origin, which lie within the cone $x^2 + y^2 \leq z^2$ (Figure 15.2).

The boundaries of the solid sphere and cone are given by the equalities $x^2 + y^2 + z^2 = 4$ and $x^2 + y^2 = z^2$ respectively. These intersect in a curve which satisfies both equations, i.e. $x^2 + y^2 + z^2 = 4 = z^2 + z^2$. Hence $z = \pm\sqrt{2}$ and $x^2 + y^2 = 2$. This is a circle of radius $\sqrt{2}$ parallel to the xy-plane, $\sqrt{2}$ units above it with centre at the point $(0, 0, \sqrt{2})$. The open set U is that portion of the cone which lies between the planes $z = 0$ and $z = \sqrt{2}$ capped by the top of the sphere (Figure 15.3). The boundary of U consists of the portion S_1 of the cone defined by $x^2 + y^2 = z^2$, $0 \leq z \leq \sqrt{2}$, and the portion of the sphere, S_2, defined by $x^2 + y^2 + z^2 = 4$, $\sqrt{2} \leq z \leq 2$. Consider the vector field

$$\mathbf{F}(x, y, z) = (xz, yz, x^2 + y^2) .$$

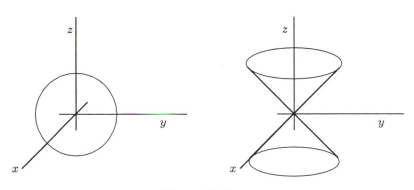

Figure 15.2.

We have div $(\mathbf{F}) = \dfrac{\partial}{\partial x}(xz) + \dfrac{\partial}{\partial y}(yz) + \dfrac{\partial}{\partial z}(x^2 + y^2) = 2z$. To evaluate

$$\iiint\limits_{U} 2z\,dxdydz$$

we use spherical polar coordinates (Example 10.3):

$$w\colon (r, \theta, \psi) \longrightarrow (r \sin\theta \cos\psi, r \sin\theta \sin\psi, r \cos\theta)\,.$$

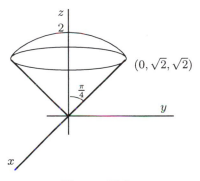

Figure 15.3.

From Figure 15.3 the domain of w is

$$0 < \theta < \frac{\pi}{4}, \quad 0 < r < 2, \quad 0 < \psi < 2\pi.$$

Since the rows of

$$w' = \begin{pmatrix} w_r \\ w_\theta \\ w_\psi \end{pmatrix} = \begin{pmatrix} \sin\theta\cos\psi & \sin\theta\sin\psi & \cos\theta \\ r\cos\theta\cos\psi & r\cos\theta\sin\psi & -r\sin\theta \\ -r\sin\theta\sin\psi & r\sin\theta\cos\psi & 0 \end{pmatrix}$$

are easily seen to be mutually orthogonal

$$\det(w') = 1 \cdot r \cdot r \sin\theta = r^2 \sin\theta \, .$$

Hence

$$\iiint\limits_U 2z dx dy dz \;\; = \;\; 2 \int_0^{\pi/4} \int_0^2 \int_0^{2\pi} r^2 \sin\theta \cdot r \cos\theta dr d\theta d\psi$$

$$= \;\; \int_0^{\pi/4} \sin 2\theta d\theta \cdot \int_0^2 r^3 dr \cdot \int_0^{2\pi} d\psi$$

$$= \;\; 2\pi \cdot \frac{r^4}{4}\Big|_0^2 \cdot \left(\frac{-\cos 2\theta}{2}\right)\Big|_0^{\pi/4} = 2\pi \cdot \frac{16}{4} \cdot \frac{1}{2} = 4\pi \, .$$

To verify the divergence theorem we must show

$$\iint\limits_{\mathbf{S_1}} \mathbf{F} + \iint\limits_{\mathbf{S_2}} \mathbf{F} = 4\pi$$

where $\mathbf{S_1}$ and $\mathbf{S_2}$ both are oriented outwards. To ensure that we do not take an incorrect sign we use the formula

$$\iint\limits_{\mathbf{S}} \mathbf{F} = \iint\limits_S (\mathbf{F} \cdot \mathbf{n})\sqrt{EG - F^2}$$

to evaluate the surface integrals where $\mathbf{n}$ is the outer normal. Since S_1 is part of the level set $x^2 + y^2 - z^2 = 0$ its unit normals are

$$\pm\frac{(2x, 2y, -2z)}{\left((2x)^2 + (2y)^2 + (2z)^2\right)^{1/2}} = \pm\frac{(x, y, -z)}{\sqrt{2}z} \, .$$

On S_1, $z > 0$ and from Figure 15.3 the outer normal has a negative z-component and hence

$$\mathbf{n_1}(x, y, z) = \frac{(x, y, -z)}{\sqrt{2}z} \, .$$

Hence

$$\mathbf{F} \cdot \mathbf{n_1} = (xz, yz, x^2 + y^2) \cdot \frac{(x, y, -z)}{\sqrt{2}z} = \frac{x^2 z + y^2 z - z(x^2 + y^2)}{\sqrt{2}z} = 0$$

and

$$\iint\limits_{\mathbf{S_1}} \mathbf{F} = 0.$$

Since S_2 is part of the sphere of radius 2 centred at the origin the outer normal at (x, y, z) is $(x, y, z)/2$. Hence

$$\mathbf{F} \cdot \mathbf{n_2} = (xz, yz, x^2 + y^2) \cdot \frac{(x, y, z)}{2} = \frac{x^2 z + y^2 z + (x^2 + y^2)z}{2} = (x^2 + y^2)z.$$

We parametrise the sphere using spherical polar coordinates (Example 10.3) and obtain

$$g: (\theta, \psi) \longrightarrow (2 \sin \theta \cos \psi, 2 \sin \theta \sin \psi, 2 \cos \theta).$$

From Figure 15.3, the domain of g is $0 < \theta < \pi/4$, $0 < \psi < 2\pi$. We have already seen in Example 11.3 that

$$\| g_\theta \times g_\psi \| = \sqrt{EG - F^2} = 2^2 \sin \theta$$

and from the above

$$\mathbf{F} \cdot \mathbf{n_2} = 4 \sin^2 \theta \cdot 2 \cos \theta = 8 \sin^2 \theta \cos \theta.$$

Hence

$$\iint_{S_2} \mathbf{F} = \int_0^{\pi/4} \int_0^{2\pi} 32 \sin^3 \theta \cos \theta \, d\theta \, d\psi.$$

Let $u = \sin \theta$. $du = \cos \theta \, d\theta$ and

$$\iint_{S_2} \mathbf{F} = 64\pi \int_0^{1/\sqrt{2}} u^3 \, du$$

$$= 64\pi \cdot \left. \frac{u^4}{4} \right|_0^{1/\sqrt{2}} = \frac{64\pi}{4} \cdot \frac{1}{4} = 4\pi$$

and

$$\iiint_U \operatorname{div}(\mathbf{F}) = \iint_{S_1} \mathbf{F} + \iint_{S_2} \mathbf{F}.$$

We have this verified the divergence theorem.

Hopefully we have developed, in the last few chapters, certain skills in handling diagrams, parametrizations and linear algebra and seen that problems in integration theory are handled by a mixture of techniques and ideas. With some practice readers will find their own preferred approach and, indeed, recognise that there are alternative approaches to a number of problems. Our next example is similar but somewhat more complicated than the previous example. We could take the same approach but instead use a few different ideas (some call them tricks) which may be useful elsewhere. This example also highlights the relationship between the change of variables for triple integrals and parametrizations of the boundary surfaces. Recall that we obtained many of the change of variables formulae for solids by "filling in" parametrization of boundary surfaces (Example 14.3). In the next example we see that restricting parametrizations of solids to the boundaries leads to parametrizations of surfaces.

Example 15.4

In this example we verify the divergence theorem on a part of a solid torus (Example 11.1). Consider the region A in the yz-plane determined by the inequalities $(y - b)^2 + z^2 \leq a$ and $b - y \leq z \leq y - b$ where $b > a > 0$. The boundary of A consists of an arc of a circle of radius a with centre $(b, 0)$ and two straight lines which pass through the centre of the circle. These are easily sketched, Figure 15.4, and we will constantly refer to this simple diagram as it contains a great deal of information.

If we revolve A about the z-axis we obtain a wedge-shaped portion U of the solid torus (see Figure 11.4 for a sketch of the torus). We parametrise the solid torus by filling in the toroidal polar coordinates given in Example 11.1 and use Figure 15.4 to find the domain of the parametrization P of the wedge U.

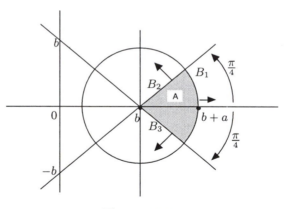

Figure 15.4.

We obtain

$$P : (r, \theta, \psi) \longrightarrow \Big((b + r\cos\theta)\cos\psi, (b + r\cos\theta)\sin\psi, r\sin\theta \Big)$$

where $0 < r < a$, $-\pi/4 < \theta < \pi/4$ and $0 < \psi < 2\pi$. Consider the vector field

$$\mathbf{F}(x, y, z) = \Big(x((x^2 + y^2)^{1/2} - b), y((x^2 + y^2)^{1/2} - b), z(x^2 + y^2)^{1/2} \Big).$$

In Cartesian coordinates,

$$
\begin{aligned}
\text{div } (\mathbf{F}) \;=\;& (x^2+y^2)^{1/2} - b + x \cdot \frac{1}{2} \cdot 2x(x^2+y^2)^{-1/2} + (x^2+y^2)^{1/2} - b \\
& + y \cdot \frac{1}{2} \cdot 2y(x^2+y^2)^{-1/2} + (x^2+y^2)^{1/2} \\
=\;& 3(x^2+y^2)^{1/2} - 2b + (x^2+y^2)\cdot(x^2+y^2)^{-1/2} \\
=\;& 4(x^2+y^2)^{1/2} - 2b
\end{aligned}
$$

and, in toroidal polar coordinates,

$$
\begin{aligned}
\text{div } (\mathbf{F}) \;=\;& 4\Big((b+r\cos\theta)^2 \cos^2\psi + (b+r\cos\theta)^2 \sin^2\psi \Big)^{1/2} - 2b \\
=\;& 4(b+r\cos\theta) - 2b = 2(b+2r\cos\theta) \ .
\end{aligned}
$$

Since

$$
P' = (P_r, P_\theta, P_\psi) = \begin{pmatrix} \cos\theta\cos\psi & -r\sin\theta\cos\psi & -(b+r\cos\theta)\sin\psi \\ \cos\theta\sin\psi & -r\sin\theta\sin\psi & (b+r\cos\theta)\cos\psi \\ \sin\theta & r\cos\theta & 0 \end{pmatrix}
$$

and P_r, P_θ and P_ψ are easily seen, by inspection, to be mutually perpendicular we have

$$
|\det(P')| = 1 \cdot r \cdot (b + r\cos\theta) \ .
$$

Hence

$$
\begin{aligned}
\iiint\limits_{U} \text{div}(\mathbf{F}) \;=\;& \int_{\substack{0 \\ (r)}}^{a} \int_{\substack{-\pi/4 \\ (\theta)}}^{\pi/4} \int_{\substack{0 \\ (\psi)}}^{2\pi} 2(b+2r\cos\theta)r(b+r\cos\theta)drd\theta d\psi \\
=\;& 4\pi \int_{-\pi/4}^{\pi/4} \int_0^a (b^2 r + 3r^2 b\cos\theta + 2r^3\cos^2\theta)drd\theta \\
=\;& 4\pi \int_{-\pi/4}^{\pi/4} \Big(\frac{b^2 a^2}{2} + ba^3\cos\theta + \frac{a^4}{2}\cdot\frac{1+\cos 2\theta}{2} \Big) d\theta \\
=\;& \frac{4\pi b^2 a^2}{2}\cdot\frac{\pi}{2} + 4\pi ba^3 (\sin\theta)\Big|_{-\pi/4}^{\pi/4} + \frac{4\pi a^4}{4}\Big(\theta + \frac{\sin 2\theta}{2}\Big)\Big|_{-\pi/4}^{\pi/4} \\
=\;& \pi^2 b^2 a^2 + 4\sqrt{2}\pi ba^3 + \frac{\pi^2 a^4}{2} + \pi a^4 \ .
\end{aligned}
$$

To verify the divergence theorem we calculate the integral over the boundary of U oriented outwards. The boundary of U is obtained by revolving the boundary of A about the z-axis. From Figure 15.4 we see that the boundary of A consists of an arc B_1 of a circle which revolves into a part of the torus and two straight lines, B_2 and B_3, which revolve into portions of cones. We denote by $\mathbf{S}_i$ the outwardly oriented surface obtained by revolving B_i about the z-axis, $i = 1, 2, 3$. We could proceed to parametrise each of these surfaces using the methods developed earlier, but instead we obtain our parametrizations by *appropriate restrictions* of P. For example B_1 is the boundary of A obtained by letting $r = a$ in the parametrization $(r, \theta) \to (b + r\cos\theta, r\sin\theta)$ of A and if we let $r = a$ in $P(r, \theta, \psi)$ we obtain a parametrization f of S_1 given by

$$f(\theta, \psi) = \big((b + a\cos\theta)\cos\psi, (b + a\cos\theta)\sin\psi, a\sin\theta\big)$$

where $-\pi/4 < \theta < \pi/4$ and $0 < \psi < 2\pi$. Since partial derivatives are calculated by fixing all except one of the variables we have $f_\theta(\theta, \psi) = P_\theta(a, \theta, \psi)$ and $f_\psi(\theta, \psi) = P_\psi(a, \theta, \psi)$. Thus the partial derivatives of f are the final two columns of P' and as the columns of P' are mutually perpendicular the first column is parallel to the normal of S_1. Since the first column of P' is a unit vector it follows that the unit normals to S_1 are

$$\pm(\cos\theta\cos\psi, \cos\theta\sin\psi, \sin\theta).$$

If $\theta = 0$ and $\psi = \pi/2$ we get the point $(0, b + a, 0)$ in S_1 with unit normals at this point given by $\pm(0, 1, 0)$. From Figure 15.4 the outward normal is $(0, 1, 0)$. Hence the outward normal at any point of $\mathbf{S_1}$ is given by $(\cos\theta\cos\psi, \cos\theta\sin\psi, \sin\theta)$. Again, using orthogonality of the columns of P we see that

$$\|f_\theta \times f_\psi\| = \|f_\theta\| \cdot \|f_\psi\| = a(b + a\cos\theta).$$

On S_1, $x^2 + y^2 = (b + a\cos\theta)^2\cos^2\psi + (b + a\cos\theta)^2\sin^2\psi = (b + a\cos\theta)^2$ and

$$\begin{aligned}\mathbf{F} &= \big(a(b + a\cos\theta)\cos\psi\cos\theta, a(b + a\cos\theta)\sin\psi\cos\theta, a\sin\theta(b + a\cos\theta)\big)\\ &= a(b + a\cos\theta)(\cos\psi\cos\theta, \sin\psi\cos\theta, \sin\theta)\\ &= a(b + a\cos\theta)\mathbf{n}.\end{aligned}$$

Hence

$$\begin{aligned}\iint_{\mathbf{S_1}} \mathbf{F} &= \iint_{S_1} \mathbf{F} \cdot \mathbf{n}\|f_\theta \times f_\psi\| d\theta d\psi\\ &= \int_0^{2\pi}\int_{-\pi/4}^{\pi/4} a(b + a\cos\theta)\mathbf{n} \cdot a(b + a\cos\theta)\mathbf{n}d\psi d\theta\\ &= 2\pi a^2 \int_{-\pi/4}^{\pi/4} (b^2 + 2ab\cos\theta + a^2\cos^2\theta)d\theta\end{aligned}$$

$$= 2\pi a^2 \cdot \frac{b^2\pi}{2} + 4\pi a^3 b(\sin\theta) \Big|_{-\pi/4}^{\pi/4} + 2\pi a^4 \int_{-\pi/4}^{\pi/4} \frac{1+\cos 2\theta}{2} d\theta$$

$$= \pi^2 a^2 b^2 + 4\pi a^3 b\sqrt{2} + \pi a^4 \frac{\pi}{2} + \frac{2\pi a^4}{2}\Big(\frac{\cos 2\theta}{2}\Big) \Big|_{-\pi/4}^{\pi/4}$$

$$= \pi^2 a^2 b^2 + 4\sqrt{2}\pi a^3 b + \frac{\pi^2 a^4}{2} + \pi a^4 .$$

The boundary B_2 of A is obtained by letting $\theta = \pi/4$ in the parametrization $(r,\theta) \to (b+r\cos\theta, r\sin\theta)$ of A and hence a parametrization of S_2 is obtained by letting $\theta = \pi/4$ in $P(r,\theta,\psi)$. This gives the parametrization

$$g(r,\psi) = \Big(\big(b+\frac{r}{\sqrt{2}}\big)\cos\psi, \big(b+\frac{r}{\sqrt{2}}\big)\sin\psi, \frac{r}{\sqrt{2}}\Big)$$

with domain $0 \le \psi \le 2\pi$, $0 \le r \le a$. As before the partial derivatives of g are obtained from the first and third columns of P', with $\theta = \pi/4$, and again, using orthogonality of the rows of P', we see that the unit normals to S_2 are obtained by normalising the second column of P'. The normals are $\pm(\cos\psi, \sin\psi, -1)/\sqrt{2}$. From Figure 15.4 the outer normal has a positive final coordinate and thus the outer normal at the point $g(r,\psi)$ is $(-\sin\psi, -\cos\psi, 1)/\sqrt{2}$. Figure 15.4 also shows that S_2 is part of the cone $x^2 + y^2 = (z+b)^2$. On S_2,

$$(x^2+y^2)^{1/2} - b = b + \frac{r}{\sqrt{2}} - b = \frac{r}{\sqrt{2}} .$$

Hence

$$\mathbf{F}\big(g(r,\psi)\big) = \Big(\big(b+\frac{r}{\sqrt{2}}\big)\cos\psi \cdot \frac{r}{\sqrt{2}}, \big(b+\frac{r}{\sqrt{2}}\big)\sin\psi \cdot \frac{r}{\sqrt{2}}, \frac{r}{\sqrt{2}}\big(b+\frac{r}{\sqrt{2}}\big)\Big)$$

$$= \big(b+\frac{r}{\sqrt{2}}\big) \cdot \frac{r}{\sqrt{2}}(\cos\psi, \sin\psi, 1) .$$

Hence

$$\mathbf{F}\cdot\mathbf{n} = \big(b+\frac{r}{\sqrt{2}}\big) \cdot \frac{r}{\sqrt{2}} \cdot \frac{1}{\sqrt{2}}(-\sin^2\psi - \cos^2\psi + 1) = 0$$

and

$$\iint_{\mathbf{S_2}} \mathbf{F} = 0.$$

The surface $\mathbf{S_3}$ is obtained by letting $\theta = -\pi/4$ and we obtain in the same way a parametrization

$$h\colon (r,\psi) \longrightarrow \Big(\big(b+\frac{r}{\sqrt{2}}\big)\cos\psi, \big(b+\frac{r}{\sqrt{2}}\big)\sin\psi, -\frac{r}{\sqrt{2}}\Big)$$

with outer normal $\left(-\dfrac{\cos \psi}{\sqrt{2}}, -\dfrac{\sin \psi}{\sqrt{2}}, -\dfrac{1}{\sqrt{2}}\right)$ at the point $h(r, \psi)$. It is easily checked that

$$\mathbf{F}\big(h(r, \psi)\big) = \left(b + \frac{r}{\sqrt{2}}\right) \cdot \frac{r}{\sqrt{2}} (\cos \psi, \sin \psi, -1),$$

and

$$\mathbf{F} \cdot \mathbf{n} = \left(b + \frac{r}{\sqrt{2}}\right) \frac{r}{\sqrt{2}} \cdot \frac{1}{\sqrt{2}} (-\sin^2 \psi - \cos^2 \psi + 1) = 0 \, .$$

Hence $\iint\limits_{S_3} \mathbf{F} = 0$. We have thus shown

$$\iiint\limits_{U} \operatorname{div}(\mathbf{F}) = \iint\limits_{S_1} \mathbf{F} + \iint\limits_{S_2} \mathbf{F} + \iint\limits_{S_3} \mathbf{F}$$

and verified the divergence theorem.

This completes our programme on integration theory. We have concentrated on motivating the basic definitions, the concept of orientation, the main theorems (Stokes' theorem and the divergence theorem) and geometrical interpretations. We have neglected mentioning practical and theoretical applications of several-variable integration theory. The methods we have covered are not trivial and need time to be mastered. We urge the reader to have patience and to keep revising until they appear obvious. Further studies, of a pure and applied nature should then be relatively easy and highly rewarding. The remaining chapters of this book are devoted to the geometry of surfaces in $\mathbb{R}^3$. The examples of surfaces that arose in integration theory are both concrete and representative and should be frequently referred to as a means of appreciating and understanding the abstract concepts we meet in the final three chapters.

EXERCISES

15.1 Prove the divergence theorem for the domain

$$U = \{(x, y, z) : 0 \leq x \leq 1, \ 0 \leq y \leq 1, \ 0 \leq z \leq 1\}.$$

15.2 Use the divergence theorem to evaluate

$$\iint\limits_{S} \mathbf{F}$$

where $\mathbf{F}(x, y, z) = (xy^2, x^2 y, y)$ and $\mathbf{S}$ is the surface of the cylinder $x^2 + y^2 \leq 3$ bounded by the planes $x + z = 0$ and $z = 0$ oriented outwards.

15.3 Evaluate

$$\iiint_U x\,dx\,dy\,dz$$

where U is the tetrahedron bounded by $x \geq 0$, $y \geq 0$, $z \geq 0$ and the plane $\dfrac{x}{a} + \dfrac{y}{b} + \dfrac{z}{c} = 1$.

15.4 Evaluate directly and also using the divergence theorem

$$\iiint_{0 \leq x,y,z \leq 1} yz^2 e^{-xyz}\,dx\,dy\,dz.$$

15.5 Use the divergence theorem to prove

$$\iint_S (x^2, -2xy, 3xz) = 3\pi$$

where $\mathbf{S}$ is the outwardly oriented boundary of the region

$$\{(x, y, z) : x \geq 0,\ y \geq 0,\ z \geq 0,\ x^2 + y^2 + z^2 \leq 4\}.$$

15.6 Let S denote a closed surface which bounds the open set V in $\mathbb{R}^3$ and let $\mathbf{n}$ denote the outward normal to S. If f and g have continuous first- and second-order partial derivatives and $\nabla^2 f = \dfrac{\partial^2 f}{\partial x^2} + \dfrac{\partial^2 f}{\partial y^2} + \dfrac{\partial^2 f}{\partial z^2}$ show that

$$\iint_S \left(f\frac{\partial g}{\partial \mathbf{n}} - g\frac{\partial f}{\partial \mathbf{n}} \right) = \iiint_V (f\nabla^2 g - g\nabla^2 f)\,dx\,dy\,dz.$$

16

Geometry of Surfaces in $\mathbb{R}^3$

Summary. *Using normal sections we define normal curvature, principal curvatures and Gaussian curvature. Geometric interpretations and a method of calculating the Gaussian curvature using parametrization are given.*

Our geometric study of surfaces in $\mathbb{R}^3$ is motivated by some very simple basic questions such as; what is curvature and how does one measure it? Is there any relationship between surface area and curvature? What is the shortest distance between two points on a surface? We start by taking an intuitive and non-rigorous look at an apparently very special case and this leads us to mathematical concepts which are both useful and natural. The surface we study is one with which we are already familiar and this simple example gives us *everything*. We have already used *all* the techniques that we require, and *all* the facts that we need are known to us – we just have to look at things in a slightly different way. The surface S we consider is the graph of the smooth function f where $f(x, y)$ is the height above sea level. We suppose f has a local maximum at (x_0, y_0) and we study S near $p = (x_0, y_0, f(x_0, y_0))$. Since the point (x_0, y_0) is a critical point of f the tangent plane to S at p is the plane through the point $(x_0, y_0, f(x_0, y_0))$ *parallel* to the xy-plane in $\mathbb{R}^3$. The unit normals to the surface at p are $\pm(0, 0, 1)$ and, for the sake of convenience, we *choose* $(0, 0, 1)$ as our unit normal $\mathbf{n}(p)$. The tangent plane of S at p consists of all vectors of the form $(v_1, v_2, 0)$.

If our notion of curvature is meaningful it should say something when we take cross-sections of a surface. For instance, if we keep getting circles when we take cross-sections we should not be surprised if the surface is a sphere and if each cross-section is either a line or a plane we expect the surface to be a plane. We consider cross-sections of $\mathbb{R}^3$ through the point p which contain the unit normal at p, $\mathbf{n}(p)$. Since cross-sections of $\mathbb{R}^3$ are two-dimensional this will cut the tangent plane and we can find a unit tangent vector at p, $\mathbf{v}$, such that our cross-section has the form

$$p + \{x\mathbf{v} + y\mathbf{n}(p) : x, y \in \mathbb{R}^2\}.$$

We identify this cross-section of $\mathbb{R}^3$ with $\mathbb{R}^2_{(x,y)}$ by the correspondence

$$p + x\mathbf{v} + y\mathbf{n}(p) \in \mathbb{R}^3 \longleftrightarrow (x, y) \in \mathbb{R}^2.$$

The intersection of this cross-section of $\mathbb{R}^3$ and the surface S is a *curve* on the surface called a *normal section* of the surface at p (Figure 16.1).

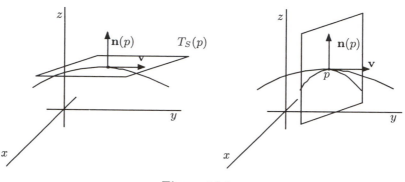

Figure 16.1.

Using the above correspondence we may identify the normal section of the surface with a curve Γ in $\mathbb{R}^2$ passing through the origin. We direct Γ by requiring $(0, 1)$ to be the unit normal at the origin and calculate its curvature using the concept of plane curvature in Chapter 7. The curvature of Γ at the origin in $\mathbb{R}^2$ will depend on the unit vector $\mathbf{v}$ and the choice of normal and we denote it by $k_p(\mathbf{v})$. We call $k_p(\mathbf{v})$ the *normal curvature* of S at p in the direction $\mathbf{v}$. By examining $k_p(\mathbf{v})$ as $\mathbf{v}$ ranges over all unit tangent vectors to S at p we hope to draw conclusions on the shape of the surface near p. For example, if $k_p(\mathbf{v}) = 1/r$ for all $\mathbf{v}$ then we might expect the surface to approximate a sphere of radius r *near* p. All normal sections are paths on the surface leading to the point p (i.e. to the top of the mountain) and the different normal curvatures distinguish between the steep and the not so steep paths. Visualising circles of different radii going through p, with p as their highest point, gives us an idea of the shape of the surface near p. Of course it is very laborious to examine all these curves and to calculate all their curvatures so instead we examine them collectively to see if any particular features of the set of *all* normal curvatures captures the essence of the shape near p.

Since f has a local maximum at (x_0, y_0) and $\mathbf{n}(p) = (0, 0, 1)$ *all* normal circles of curvature will lie on the *same* side of the tangent plane and on the opposite side to the normal. This means that all normal curvatures will be negative. Choosing $(0, 0, -1)$ as unit normal at p changes the signs of all normal curvatures. If we were examining a local minimum at (x_0, y_0) and $\mathbf{n}(p) = (0, 0, 1)$ we would have found that all normal curvatures were positive.

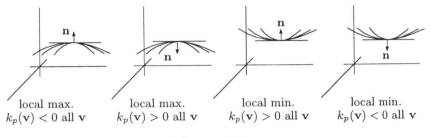

local max.	local max.	local min.	local min.
$k_p(\mathbf{v}) < 0$ all $\mathbf{v}$	$k_p(\mathbf{v}) > 0$ all $\mathbf{v}$	$k_p(\mathbf{v}) > 0$ all $\mathbf{v}$	$k_p(\mathbf{v}) < 0$ all $\mathbf{v}$

Figure 16.2.

But changing the normal or turning the surface upside down – this is equivalent to changing local maxima into local minima and conversely – does not affect the shape of the surface in any way and we conclude: if all normal curvatures are strictly positive or all are strictly negative then the shape of S near p is similar to an ellipsoid (Figure 16.2). These points are called *elliptic* points of the surface. We can extend this analysis to arbitrary critical points and clearly if, at a point p, normal curvature takes both positive and negative values then the surface near p is similar in shape to a saddle point. Such points are called *hyperbolic* points of the surface. Our preliminary investigation has shown that information on the shape of the surface can be obtained from the range of values taken by $k_p(\mathbf{v})$ as $\mathbf{v}$ varies over the set of unit tangent vectors at p. Since the range is determined by the extremal values we let $k_1(p) = \max\limits_{\|\mathbf{v}\|=1} k_p(\mathbf{v})$ and $k_2(p) = \min\limits_{\|\mathbf{v}\|=1} k_p(\mathbf{v})$ and call $k_1(p)$ and $k_2(p)$ the *principal curvatures* of the surface at p. If the maximum or minimum occurs at a vector $\mathbf{v}$ we call $\mathbf{v}$ a *principal curvature direction*.

We consider the various possibilities that may arise for $k_1(p)$ and $k_2(p)$. If $k_1(p) = k_2(p)$ then we call p an *umbilic point* and if $k_1(p) = k_2(p) = 0$ we call p a *flat spot*. If p is umbilic then $k_p(\mathbf{v}) = k_1(p)$ for all $\mathbf{v}$, $\|\mathbf{v}\| = 1$, and if $k_1(p) \neq 0$ then the surface near p is similar to a portion of a sphere of radius $1/|k_1(p)|$. If p is a flat spot then all normal curvatures are zero and the surface near p is very flat and almost like a part of a plane. At any umbilic point all directions are principal curvature directions.

We now consider non-umbilic points, i.e. $k_1(p) > k_2(p)$. If $k_2(p) > 0$ then for any $\mathbf{v}$, $\|\mathbf{v}\| = 1$, we have $0 < k_2(p) \le k_p(\mathbf{v})$ and if $k_1(p) < 0$ then $k_p(\mathbf{v}) \le k_1(p) < 0$ for all $\mathbf{v}$, $\|\mathbf{v}\| = 1$. Hence, in both cases, all normal curvatures have the same sign and near p the surface is shaped like an ellipsoid at one of its extreme points (i.e. like the surface $(x/a)^2 + (y/b)^2 + (z/c)^2 = 1$ at the point $(0, 0, c)$). If $k_2(p) < 0 < k_1(p)$ then the surface near p is shaped like a saddle point.

We summarise our conclusions in a more concise fashion by introducing the concept of *Gaussian curvature*.

Definition 16.1

The Gaussian curvature, $K(p)$, at a point p on a surface S is the product of the principal curvatures, $k_1(p)k_2(p)$, at the point.

We have noted already that $k_p(\mathbf{v})$ depends on the choice of normal – changing the normal changes the sign of $k_p(\mathbf{v})$ – but since two changes of sign cancel one another it follows that $K(p)$ *does not depend* on the choice of normal.

Although we have only considered critical points on the graph in our analysis we will see shortly that our analysis applies to *all* points on a surface and for this reason we state the following result in its full generality.

Proposition 16.1

At a non-umbilic point on a surface S in $\mathbb{R}^3$ we have:

$$K(p) > 0 \quad \Longleftrightarrow \quad \text{near } p, S \text{ is shaped like an ellipsoid,}$$
$$K(p) < 0 \quad \Longleftrightarrow \quad \text{near } p, S \text{ is shaped like a saddle point,}$$
$$K(p) = 0 \quad \Longleftrightarrow \quad \text{near } p, S \text{ is shaped like a cylinder.}$$

At an umbilic point $K(p) \ge 0$ and

$$K(p) > 0 \quad \Longleftrightarrow \quad \text{near } p, S \text{ is shaped like a sphere,}$$
$$K(p) = 0 \quad \Longleftrightarrow \quad \text{near } p, S \text{ is very flat.}$$

We discuss the case of zero Gaussian curvature in more detail later.

All our conclusions followed from taking the ordinary *plane curvatures* of certain curves on the surface at the point p. We know, however, that plane curvature only involves the *first and second derivatives* at the point and so in our analysis we can *replace f by any function g* which has the same first-

and second-order partial derivatives as f at (x_0, y_0). We suppose for simplicity $(x_0, y_0) = (0, 0)$ and $f(0, 0) = 0$. Hence $p = (0, 0, 0)$. By Taylor's theorem the simplest choice of g available, since we are looking at a critical point of f, is

$$g(x, y) = \frac{1}{2}(f_{xx}(0, 0)x^2 + 2f_{x,y}(0, 0)xy + f_{yy}(0, 0)y^2).$$

The tangent planes of the graphs of f and g coincide and the normal curvatures to both graphs are equal at $(0, 0, 0)$. If we let $A = \frac{1}{2}f_{xx}(0, 0)$, $B = \frac{1}{2}f_{xy}(0, 0)$ and $C = \frac{1}{2}f_{yy}(0, 0)$ we are reduced to studying normal curvatures of the graph of

$$g(x, y) = Ax^2 + 2Bxy + Cy^2$$

at the origin in $\mathbb{R}^3$. The tangent plane at $p = (0, 0, 0)$ is the xy-plane and consists of the points $\{(v_1, v_2, 0) : v_1, v_2 \in \mathbb{R}\}$, and $\mathbf{n} = \mathbf{n}(p) = (0, 0, 1)$ is the unit normal. Let $\mathbf{v} = (v_1, v_2, 0)$ be a fixed unit vector. The section of $\mathbb{R}^3$ determined by $\mathbf{n}$ and $\mathbf{v}$ is the plane

$$\{(tv_1, tv_2, s) : t, s \in \mathbb{R}\} \tag{16.1}$$

and the graph of g is the set

$$\{(x, y, Ax^2 + 2Bxy + Cy^2) : x, y \in \mathbb{R}\} \tag{16.2}$$

The points which lie in (16.1) and (16.2) are the normal section of the surface through p defined by $\mathbf{v}$ and are easily seen to have the form

$$\{(tv_1, tv_2, (Av_1^2 + 2Bv_1v_2 + Cv_2^2)t^2), t \in \mathbb{R}\}$$
$$= \{t\mathbf{v} + t^2(Av_1^2 + 2Bv_1v_2 + Cv_2^2)\mathbf{n} : t \in \mathbb{R}\}. \tag{16.3}$$

From our identification of the plane in $\mathbb{R}^3$ spanned by $\mathbf{v}$ and $\mathbf{n}$ with $\mathbb{R}^2_{(x,y)}$ we see that the normal section is the set

$$\{(t, (Av_1^2 + 2Bv_1v_2 + Cv_2^2)t^2) : t \in \mathbb{R}\}$$

and may be identified with the graph of the function $h \colon \mathbb{R} \to \mathbb{R}$

$$h(t) = (Av_1^2 + 2Bv_1v_2 + Cv_2^2)t^2.$$

If we parametrise the graph of h by $t \to (t, h(t))$ we obtain $(1, 0)$ as unit tangent at $(0, 0)$ and $(0, 1)$ as unit normal (see chapter 7). Hence the curvature of the graph at the origin, directed by this parametrization, is the normal curvature of S at p in the direction $\mathbf{v}$. In Example 7.1 we showed that the curvature of the graph of h at $(t, h(t))$ is

$$\frac{h''(t)}{(1 + h'(t)^2)^{3/2}}.$$

At $t = 0$ we get

$$\frac{h''(0)}{(1 + h'(0)^2)^{3/2}} = 2(Av_1^2 + 2Bv_1v_2 + Cv_2^2).$$

Returning now to the function f, via h and g, we have shown

$$k_p(\mathbf{v}) = f_{xx}(0,0)v_1^2 + 2f_{xy}(0,0)v_1v_2 + f_{y,y}(0,0)v_2^2 = \mathbf{v}H_{f(0,0)}{}^t\mathbf{v}$$

where

$$H_{f(0,0)} = \begin{pmatrix} f_{xx}(0,0) & f_{xy}(0,0) \\ f_{xy}(0,0) & f_{yy}(0,0) \end{pmatrix}$$

is the *Hessian* of f at $(0,0)$. (We have taken the liberty of using $\mathbf{v}$ to denote both (v_1, v_2) and $(v_1, v_2, 0)$ and hope that this does not cause any confusion – it is a practice that we do not recommend except in special circumstances.)

We are now in familiar territory and can use our knowledge of the function $\mathbf{v}H_{f(0,0)}{}^t\mathbf{v}$ as $\mathbf{v}$ ranges over all unit vectors. In Chapter 4 we showed that the maximum and minimum values of $\mathbf{v}H_{f(0,0)}{}^t\mathbf{v}$ over $\|\mathbf{v}\| = 1$ are the *eigenvalues* of $H_{f(0,0)}$ and are achieved at the corresponding eigenvectors. Hence the eigenvectors of $H_f(0,0)$ are the principal curvature directions and p is non-umbilic if and only if $H_f(0,0)$ has two (distinct) eigenvalues. Since vectors corresponding to different eigenvalues are perpendicular (Exercise 4.9) it follows that at a non-umbilic point there exist *precisely two* principal curvature directions which are perpendicular to one another – hence if we know one we can easily find the other. Note that we do not distinguish between the directions $\mathbf{v}$ and $-\mathbf{v}$.

We now discuss the case previously omitted, i.e. $K(p) = 0$, which corresponds to a degenerate critical point of the function f. If $K(p) = 0$ and p is umbilic then since $K(p) = k_1(p)k_2(p)$ and $k_1(p) = k_2(p)$ it follows that $k_1(p) = k_2(p) = 0$ and p is a flat spot and we have already considered this case. If $K(p) = 0$ and p is not umbilic then we have two possibilities, $k_2(p) < 0 = k_1(p)$ and $k_2(p) = 0 < k_1(p)$, and which occurs depends on the choice of normal. Since both have the same geometrical interpretation we just consider the first one. If $\mathbf{v}$ is the principal curvature direction associated with $k_1(p)$, i.e. $k_\mathbf{v}(p) = k_1(p)$, then the normal section near p in the direction $\mathbf{v}$ is approximately a straight line. If $\mathbf{w} \perp \mathbf{v} = 0$ then $\mathbf{w}$ is the other principal curvature direction and the normal section in this direction is approximately a circle of radius $1/|k_2(p)|$ on the opposite side of the surface to the normal (Figures 16.3(a) and (b)).

Figure 16.3(b) suggests a *cylinder* as an example and indeed a cylinder has Gaussian curvature zero at *all* points. For the cylinder in Figure 16.3(c) we have $k_1(p) = 0$ and $k_2(p) = -1/r$. In general, if $K(p) = 0$ then all normal circles of curvature will lie on the same side of the tangent plane and all normal

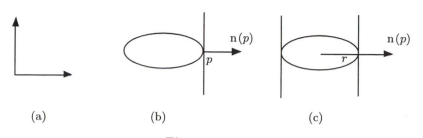

(a)   (b)   (c)

Figure 16.3.

curvatures will have the same sign, that is, either all non-negative or all non-positive. The cylinder is a good typical example of a surface with $K(p) = 0$. One should, however, not assume that every surface with $K(p) = 0$ is of this type as one can find quite strange surfaces with $K(p) = 0$ at isolated points. Our geometric interpretations are meant as a rough guide and as such are reasonably useful in visualising the shape of the surface but do not, of course, explain the full subtlety of many situations – this requires further analysis.

So far we have considered a rather special situation – a critical point on a surface which is the graph of a function – and it is time to see how representative this is of the general situation. By the implicit function theorem, every surface is locally the graph of a function – if we wished, we could even take this as the definition of a surface – and since the concepts we have introduced are all *local* properties of the surface there is no loss of generality in looking at graphs.

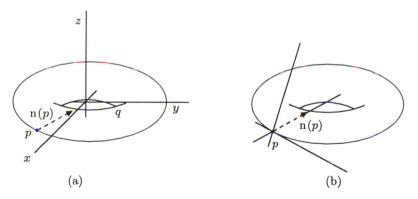

(a)   (b)

Figure 16.4.

What about critical points? Well, if looked at the right way, *every point is a critical point*. For example, consider the torus in $\mathbb{R}^3$ and take p as a typical point on the surface (Figure 16.4(a)). To turn p into a critical point we must choose a new coordinate system for $\mathbb{R}^3$. We choose our first two coordinates so that the tangent plane to the surface at p corresponds to the (x, y)-plane and then take the z-direction as one of the normal directions (Figure 16.4(b)).

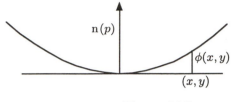

Figure 16.5.

We define a function ϕ on the tangent plane near p by letting $\phi(x, y)$ denote the distance from the tangent plane in the $\mathbf{n}(p)$ direction to the surface (Figure 16.5). This means that the surface S near p is the graph of ϕ and that ϕ, defined on the tangent space near p, has a local minimum at p. One can carry out a similar analysis at any point on the surface, although it is not so easy to sketch at points which become saddle points, e.g. the point q in Figure 16.4(a).

Thus we conclude that our analysis covers all points and we have defined normal curvatures, principal curvatures and Gaussian curvature *everywhere* on a surface. This also shows that we may continue our analysis *in full generality* by turning again to critical points on a graph.

Gaussian curvature is the most important number that we can geometrically associate with a point on a surface. To use it effectively we need to be able to calculate it directly from any parametrization, without first finding normal curvatures or principal curvatures. To find a way to do this we turn again to normal curvatures and use results from Chapter 7 on the curvature of plane curves.

If P is a unit speed parametrised curve in $\mathbb{R}^2$ then

$$P'(t) = T(t), \qquad T'(t) = \kappa(t)N(t), \qquad \kappa(t) = \langle T'(t), N(t) \rangle.$$

Since $\langle T(t), N(t) \rangle = 0$ we get on differentiating that

$$\kappa(t) = -\langle N'(t), T(t) \rangle. \tag{16.4}$$

The unit vector $\mathbf{v}$ is a unit tangent to the normal section of S, $\Gamma_{(p,\mathbf{v})}$, defined by $\mathbf{v}$ and, moreover, a unit normal at any point of $\Gamma_{(p,\mathbf{v})}$ is also a unit normal to the surface. Hence we can find a smooth unit normal $\mathbf{n}$ to the surface near p and a parametrization $P: [-a, a] \to \Gamma_{(p,\mathbf{v})}$ such that $P(0) = p$, $P'(0) = \mathbf{v}$ and $N(t) = \mathbf{n}(P(t))$ for all $t \in [-a, a]$. If $t \in [-a, a]$ then

$$\frac{d}{dt}\left(\mathbf{n}(P(t))\right) = N'(t). \tag{16.5}$$

The derivative in (16.5) does not depend on the parametrization as long as it goes through p with velocity $\mathbf{v}$. In fact, this exposes the true role of tangent vectors– they enable us to define directional derivatives of functions defined on the surface. If $F : S \to \mathbb{R}^n$ and $p \in S$ then the *directional derivative* of F at p in the direction of the tangent vector $\mathbf{w}$ at p is given by

$$D_{\mathbf{w}}F(p) = \frac{d}{dt}(F \circ \phi)(t)\big|_{t=0}$$

where ϕ is *any* differentiable mapping from $(-a, a)$ into S such that $\phi(0) = p$ and $\phi'(0) = \mathbf{w}$.

Hence (16.5) can be rewritten as $D_{\mathbf{v}}\mathbf{n}(p) = N'(0)$ and, by (16.4),

$$k_p(\mathbf{v}) = -\langle D_{\mathbf{v}}\mathbf{n}(p), \mathbf{v} \rangle$$

for any unit tangent vector $\mathbf{v} \in T_p(S)$. Since $\langle \mathbf{n}(p), \mathbf{n}(p) \rangle = 1$ the product rule for differentiation implies

$$\langle D_{\mathbf{v}}\mathbf{n}, \mathbf{n} \rangle + \langle \mathbf{n}, D_{\mathbf{v}}\mathbf{n} \rangle = 0$$

i.e. $\langle D_{\mathbf{v}}\mathbf{n}, \mathbf{n} \rangle = 0$ and $D_{\mathbf{v}}\mathbf{n}(p) \perp \mathbf{n}(p)$. Hence $D_{\mathbf{v}}\mathbf{n}(p)$ also belongs to the tangent space at p and we can define a *linear* mapping from the *tangent space* of S at p, $T_p(S)$, into itself by

$$\mathbf{v} \in T_p(S) \longrightarrow -D_{\mathbf{v}}\mathbf{n}(p) \in T_p(S).$$

This extremely important mapping is called the *Weingarten mapping* or *shape operator*. We denote it by L_p. With this notation $k_p(\mathbf{v}) = \langle L_p(\mathbf{v}), \mathbf{v} \rangle$ for every *unit tangent vector* $\mathbf{v}$ at p.

In the case where $S = \text{graph}(f)$, $p = (0, 0, 0)$, $\mathbf{n}(p) = (0, 0, 1)$ and $T_p(S) = \{(v_1, v_2, 0) : v_1, v_2 \in \mathbb{R}\}$ we note that $\mathbf{e}_1 = (1, 0, 0)$ and $\mathbf{e}_2 = (0, 1, 0)$ form an orthonormal basis for $T_p(S)$ and, from the above discussion,

$$k_p(\mathbf{v}) = \mathbf{v} H_{f(0,0)}{}^t\mathbf{v} = -\langle D_{\mathbf{v}}\mathbf{n}(p), \mathbf{v} \rangle = \langle L_p(\mathbf{v}), \mathbf{v} \rangle$$

for any unit vector $\mathbf{v} \in T_p(S)$. We recall (Example 10.1) that S is parametrised by $\phi(x, y) = (x, y, f(x, y))$. If $\mathbf{v} = (v_1, v_2, 0) \in T_p(S)$, $P(t) = \phi(tv_1, tv_2) = (tv_1, tv_2, f(tv_1, tv_2))$ then $P(0) = p$ and, by the chain rule,

$$P'(t) = (v_1, v_2, v_1 f_x(tv_1, tv_2) + v_2 f_y(tv_1, tv_2)).$$

Hence $P'(0) = (v_1, v_2, 0) = \mathbf{v}$ and

$$L_p(\mathbf{v}) = -D_{\mathbf{v}}\mathbf{n}(p) = \frac{d}{dt}\Big(\mathbf{n}(P(t))\Big)\big|_{t=0}.$$

Since

$$\mathbf{n}(\phi(x,y)) = \frac{(-f_x(x,y), -f_y(x,y), 1)}{(1 + f_x^2(x,y) + f_y^2(x,y))^{\frac{1}{2}}}$$

we have

$$L_p(\mathbf{v}) = \frac{d}{dt}\left(\frac{(-f_x(tv_1, tv_2), -f_y(tv_1, tv_2), -1)}{(1 + f_x^2(tv_1, tv_2) + f_y^2(tv_1, tv_2))^{\frac{1}{2}}}\right)\Bigg|_{t=0}$$

and after some routine but tedious calculations, involving the chain rule, we get

$$L_p(\mathbf{v}) = \big(v_1 f_{xx}(0,0) + v_2 f_{xy}(0,0),\ v_1 f_{xy}(0,0) + v_2 f_{yy}(0,0),\ 0\big).$$

In particular, $\langle L_p(\mathbf{e}_1), \mathbf{e}_1 \rangle = f_{xx}(0,0)$, $\langle L_p(\mathbf{e}_1), \mathbf{e}_2 \rangle = \langle L_p(\mathbf{e}_2), \mathbf{e}_1 \rangle = f_{xy}(0,0)$, $\langle L_p(\mathbf{e}_2), \mathbf{e}_2 \rangle = f_{yy}(0,0)$ and $H_{f(0,0)}$ is the matrix of L_p relative to the basis $\{\mathbf{e}_1, \mathbf{e}_2\}$ for $T_p(S)$. Since $H_{f(0,0)}$ is symmetric the linear operator L_p is a *self-adjoint* or *symmetric* operator and will be represented by a symmetric matrix relative to any orthonormal basis. Hence the eigenvalues and eigenvectors of L_p are the principal curvatures and principal curvature directions respectively and $\det(L_p) = K(p)$. This analysis is valid at *any* point on the surface and thus to find the Gaussian curvature it suffices to calculate the determinant of the shape operator relative to *any* orthonormal basis for the tangent space at the point. If ϕ is any parametrization of S then $\{\phi_x, \phi_y\}$ is a basis for the tangent space. By the Gram–Schmidt orthogonalization process

$$\left\{\frac{\phi_x}{\|\phi_x\|},\ \frac{\phi_y - \langle \phi_y, \phi_x \rangle \dfrac{\phi_x}{\|\phi_x\|^2}}{\left\| \phi_y - \langle \phi_y, \phi_x \rangle \dfrac{\phi_x}{\|\phi_x\|^2} \right\|}\right\}$$

is an orthonormal basis for the tangent space.

In terms of our previous notation, $E = \phi_x \cdot \phi_x$, $F = \phi_x \cdot \phi_y$ and $G = \phi_y \cdot \phi_y$, this basis can be rewritten as

$$\left\{\frac{\phi_x}{\sqrt{E}},\ \frac{E\phi_y - F\phi_x}{\sqrt{E}\sqrt{EG - F^2}}\right\}.$$

The matrix for L_p relative to this orthonormal basis is

$$\begin{pmatrix} \dfrac{L_p\phi_x \cdot \phi_x}{E} & \dfrac{EL_p\phi_y \cdot \phi_x - FL_p\phi_x \cdot \phi_x}{E\sqrt{EG - F^2}} \\[4mm] \dfrac{EL_p\phi_y \cdot \phi_x - FL_p\phi_x \cdot \phi_x}{E\sqrt{EG - F^2}} & \dfrac{E^2 L_p\phi_y \cdot \phi_y - 2EFL_p\phi_x \cdot \phi_y + F^2 L_p\phi_x \cdot \phi_x}{E(EG - F^2)} \end{pmatrix}$$

and its determinant, after some simplification, has the form

$$K(p) = \frac{\langle L\phi_x, \phi_x \rangle \langle L\phi_y, \phi_y \rangle - \langle L\phi_x, \phi_y \rangle^2}{EG - F^2}.$$

This formula works for arbitrary parametrizations but can be impractical since it requires one to first find the Weingarten mapping. The appearance of $EG - F^2$, which we previously met in calculating surface area (chapter 11), suggests some relationship between Gaussian curvature and surface area. This comes to light in the Gauss–Bonnet theorem (chapter 18). Now $\phi_x(x, y)$ is a tangent vector at $\phi(x, y)$ and if $P(t) = \phi(x + t, y)$ for all t in some interval about the origin then

$$D_{\phi_x(x,y)}(F) = \frac{d}{dt}(F \circ \phi(x + t, y))$$

for any function $F : S \rightarrow \mathbb{R}^n$. Since $\mathbf{n}(\phi(x, y)) = \frac{\phi_x(x,y) \times \phi_y(x,y)}{\|\phi_x(x,y) \times \phi_y(x,y)\|}$ and $\langle \phi_x(x, y), \mathbf{n}(\phi(x, y)) \rangle = 0$ for all (x, y) in the domain of ϕ we have

$$
\begin{aligned}
0 &= D_{\phi(x,y)}\Big(\langle \phi_x(x, y), \ \mathbf{n}(\phi(x, y)) \rangle \Big) \\
&= \frac{d}{dt} \langle \phi_x(x + t, y), \ \mathbf{n}(\phi(x + t, y)) \rangle \big|_{t=0} \\
&= \langle \frac{d}{dt} \phi_x(x + t, y) \big|_{t=0}, \ \mathbf{n}(\phi(x, y)) \rangle \\
&\qquad + \langle \phi_x(x, y), \ \frac{d}{dt} \mathbf{n}(\phi(x + t, y)) \big|_{t=0} \rangle \\
&= \langle \phi_{xx}(x, y), \ \mathbf{n}(\phi(x, y)) \rangle + \langle \phi_x(x, y), \ (D_{\phi_x(x,y)}\mathbf{n})(\phi(x, y)) \rangle \\
&= \langle \phi_{xx}(x, y), \ \mathbf{n}(\phi(x, y)) \rangle - \langle \phi_x(x, y), \ L_{\phi(x,y)}(\phi_x(x, y)) \rangle
\end{aligned}
$$

To simplify the notation we write L in place of $L_{\phi(x,y)}$, assume all derivatives of ϕ are taken at (x, y) and that $\mathbf{n}$ is evaluated at $\phi(x, y)$. The above calculation shows that $\langle L(\phi_x), \phi_x \rangle = \langle \phi_{xx}, \mathbf{n} \rangle$ and we denote this quantity by l. Similarly

$$\langle L(\phi_y), \phi_x \rangle = \langle \phi_{xy}, \mathbf{n} \rangle = m$$

and

$$\langle L(\phi_x), \phi_x \rangle = \langle \phi_{yy}, \mathbf{n} \rangle = n.$$

Putting these together we obtain the following proposition.

Proposition 16.2

The Gaussian curvature at a point p on a surface is

$$K(p) = \frac{ln - m^2}{EG - F^2}.$$

We can now calculate Gaussian curvature directly from any parametrization.

Example 16.1

We calculate the Gaussian curvature of the surface $z = xy$. This surface is the graph of the function $f(x, y) = xy$ and we obtain a parametrization ϕ by letting $\phi(x, y) = (x, y, xy)$. We have

$$\phi_x = (1, 0, y), \qquad \phi_y = (0, 1, x)$$

$$E = \phi_x . \phi_x = 1 + y^2, \quad F = \phi_x . \phi_y = xy, \quad G = \phi_y . \phi_y = 1 + x^2$$

$$EG - F^2 = (1 + y^2)(1 + x^2) - x^2 y^2 = 1 + x^2 + y^2$$

$$\phi_x \times \phi_y = \begin{vmatrix} \mathbf{i} & \mathbf{j} & \mathbf{k} \\ 1 & 0 & y \\ 0 & 1 & x \end{vmatrix} = (-y, -x, 1)$$

$$\mathbf{n} = \frac{\phi_x \times \phi_x}{\|\phi_x \times \phi_x\|} = \frac{(-y, -x, 1)}{(1 + y^2 + x^2)^{1/2}}$$

$$\phi_{xx} = (0, 0, 0), \quad \phi_{xy} = (0, 0, 1), \quad \phi_{yy} = (0, 0, 0)$$

$$l = \langle \phi_{xx}, \mathbf{n} \rangle = 0, \quad m = \langle \phi_{xy}, \mathbf{n} \rangle = \frac{1}{(1 + x^2 + y^2)^{1/2}}, \quad n = \langle \phi_{yy}, \mathbf{n} \rangle = 0.$$

Hence, if $p = (x, y, xy)$, then

$$K(p) = \frac{-m^2}{EG - F^2} = \frac{-1}{(1 + x^2 + y^2)^2}.$$

Since the Gaussian curvature is always strictly negative the surface $z = xy$ consists entirely of points which, looked at critically, are saddle points.

In this chapter we have covered a lot of theoretical and practical material in identifying and calculating the geometric concept of Gaussian curvature. In the next chapter we will summarise the information we already have on this important concept and discuss further geometric implications.

EXERCISES

16.1 Calculate the Gaussian curvature at an arbitrary point of the helicoloid parametrised by

$$P(t, \theta) = (t \cos \theta, t \sin \theta, b\theta).$$

16.2 Find the principal curvatures and the Gaussian curvature at an arbitrary point on the cone $z^2 = x^2 + y^2$.

16.3 Let S denote the surface parametrised by $\phi(u, v) = (u, v, u^2 + v^2)$, $(u, v) \in \mathbb{R}^2$. Show that the curve Γ parametrised by

$$P(t) = \phi(t^2, t), \quad -\frac{1}{2} < t < 2,$$

lies in S. Find the unit tangent to Γ at $P(1)$ and find the (absolute) normal curvature to S at $P(1)$ in the direction $P'(1)$.

16.4 If $\phi: U \to S$ is a parametrised surface and normal curvature is calculated using $\dfrac{\phi_x \times \phi_y}{\|\phi_x \times \phi_y\|}$ show that at $p = \phi(x, y)$

$$\kappa_p\left(\frac{v_1\phi_x + v_2\phi_y}{\|v_1\phi_x + v_2\phi_y\|}\right) = \frac{lv_1^2 + 2mv_1v_2 + nv_2^2}{Ev_1^2 + 2Fv_1v_2 + Gv_2^2}$$

for $(v_1, v_2) \neq (0, 0) \in \mathbb{R}^2$.

16.5 Find the elliptic and hyperbolic points on the surface parametrised by $\phi(u, v) = (u, v, u^3 + v^3)$.

16.6 Prove that the surface parametrised by

$$f(s, t) = (\cos s, 2\sin s, t), \quad 0 < s < 2\pi, \ t > 0$$

has constant Gaussian curvature.

16.7 If the surface S is defined by the equations

$$x = a(u + v), \quad y = b(u - v), \quad z = uv$$

show that the coordinate curves are straight lines.

16.8 Find the Gaussian curvature at an arbitrary point on the ellipsoid

$$\frac{x^2}{a^2} + \frac{y^2}{b^2} + \frac{z^2}{c^2} = 1.$$

16.9 Find the Gaussian curvature of the surface $z = x^2 - y^2$. Sketch this surface near the point $p = (0, 0, 0)$ and show that the normals at p are $\pm(0, 0, 1)$. Taking $\mathbf{n} = (0, 0, 1)$ examine the surface, find the principal curvatures and principal curvature directions at p. If $\mathbf{v} = (v_1, v_2, 0)$ show that

$$\kappa_p(\mathbf{v}) = v_1^2 - v_2^2.$$

Find a tangent direction with normal curvature zero and hence or otherwise find a straight line in the surface. Compare your results with Example 16.1. Why do you think they are similar?

16.10 Let
$$A = \begin{pmatrix} 6 & -4 & -2 & 2 \\ -4 & 3 & -4 & 6 \\ -2 & -4 & 6 & -2 \\ 2 & 6 & -2 & 11 \end{pmatrix} = (a_{ij})_{1 \le i,j \le 4}.$$

Find a *quadratic form*, i.e. a function $f \colon \mathbb{R}^4 \to \mathbb{R}$ of the form $f(x_1, x_2, x_3, x_4) = \sum_{i,j=1}^4 a_{ij} x_i x_j$, such that $H_{f(0,0)} = A$. Show that f has a critical point at the origin and find the nature of this critical point.

16.11 Show that a quadratic form in n variables (see Exercise 16.10) which has a strict local minimum at the origin can be written as a sum of squares. What happens if the origin is a local maximum or a saddle point? Express the function f in Exercise 16.10 as a sum of squares.

17

Gaussian Curvature

Summary. *We discuss lines of curvature, mean curvature, minimal surfaces and the role of Gaussian curvature in analysing the shape of a surface in $\mathbb{R}^3$.*

We begin by recalling the concepts we introduced and the results we obtained in the previous chapter concerning the shape of a surface near a point p. All these results were obtained using *plane curvature*. We defined or obtained the following:

(a) $k_p(\mathbf{v})$, the *normal curvature* at p in the direction of the unit tangent vector $\mathbf{v}$ at p

(b) *principal curvatures* at p, $k_1(p)$ and $k_2(p)$, where

$$k_1(p) = \max_{\|\mathbf{v}\|=1} k_p(\mathbf{v}) \quad \text{and} \quad k_2(p) = \min_{\|\mathbf{v}\|=1} k_p(\mathbf{v})$$

(c) *principal curvature directions*, i.e. tangent vectors $\mathbf{v}_1$ and $\mathbf{v}_2$ such that

$$k_p(\mathbf{v}_i) = k_i(p) \quad \text{for } i = 1, 2$$

(d) *umbilic points*, i.e. points where $k_1(p) = k_2(p)$, and *flat spots*, i.e. where $k_1(p) = k_2(p) = 0$

(e) at an *umbilic* point all (tangential) directions are principal curvature directions; at a *non-umbilic* point there are precisely two principal curvature directions $\mathbf{v}_1$ and $\mathbf{v}_2$ which are *perpendicular* to one another

213

(f) *Gaussian curvature* at p, $K(p)$, defined as

$$K(p) = k_1(p)k_2(p)$$

(g) the *Weingarten mapping* or *shape operator*

$$L_p: T_p(S) \to T_p(S), \quad L_p(\mathbf{v}) = -\left(D_{\mathbf{v}}\mathbf{n}\right)(p)$$

where $\mathbf{n}$ is a smooth unit normal defined near p

(h) the formulae

$$k_p(\mathbf{v}) = -\langle D_{\mathbf{v}}\mathbf{n}(p), \mathbf{v}\rangle = \langle L_p(\mathbf{v}), \mathbf{v}\rangle$$

$$K(p) = \det(L_p) = \frac{ln - m^2}{EG - F^2}$$

where ϕ is any parametrization,

$$E = \phi_x \cdot \phi_x, \quad F = \phi_x \cdot \phi_y, \quad G = \phi_y \cdot \phi_y$$

$$l = \phi_{xx} \cdot \mathbf{n}, \quad m = \phi_{xy} \cdot \mathbf{n}, \quad n = \phi_{yy} \cdot \mathbf{n}$$

$$\mathbf{n} = \frac{\phi_x \times \phi_y}{\|\phi_x \times \phi_y\|}.$$

We continue our investigations. If p is umbilic then all normal curvatures are equal. If p is *non-umbilic* and k_1, k_2, $\mathbf{v}_1$, and $\mathbf{v}_2$ are the principal curvatures and the corresponding principal curvature directions then any *unit tangent vector* $\mathbf{v}$ at p has the form

$$\mathbf{v} = \cos\theta\mathbf{v}_1 + \sin\theta\mathbf{v}_2$$

for some real number θ. Since $\mathbf{v}_1$ and $\mathbf{v}_2$ are principal curvature directions, they are *eigenvectors* of the operator L_p, i.e.

$$L_p(\mathbf{v}_1) = k_1(p)\mathbf{v}_1 \quad \text{and} \quad L_p(\mathbf{v}_2) = k_2(p)\mathbf{v}_2$$

and, as p is non-umbilic, $\langle \mathbf{v}_1, \mathbf{v}_2\rangle = 0$. By (h),

$$\begin{aligned}
k_p(\mathbf{v}) = \langle L_p(\mathbf{v}), \mathbf{v}\rangle &= \langle L_p(\cos\theta\mathbf{v}_1 + \sin\theta\mathbf{v}_2), \cos\theta\mathbf{v}_1 + \sin\theta\mathbf{v}_2\rangle \\
&= \langle \cos\theta L_p(\mathbf{v}_1) + \sin\theta L_p(\mathbf{v}_2), \cos\theta\mathbf{v}_1 + \sin\theta\mathbf{v}_2\rangle \\
&= \cos^2\theta k_1(p) + \sin^2\theta k_2(p).
\end{aligned}$$

This is known as *Euler's formula* and shows that normal curvature in any direction can be recovered from the principal curvatures and the principal curvature directions.

The Gaussian curvature contains less information than the principal curvatures, that is to say if we know the principal curvatures then we can calculate the Gaussian curvature but from the Gaussian curvature alone we cannot calculate the principal curvatures. Thus, at first glance, it appears that in using Gaussian curvature we may be neglecting important information.

However, experience and subsequent results show that the information lost is generously compensated by other gains. To begin with, Gaussian curvature is a single real number assigned to each point on a surface – the principal curvatures and directions involve two real numbers and two vectors and the Weingarten mapping involves a 2×2 symmetric matrix. Thus Gaussian curvature has the advantage of *simplicity*. We have already seen that Gaussian curvature may be easily calculated from any parametrization whereas it may be difficult to calculate the Weingarten mapping or the principal curvatures. The Weingarten mapping and the principal curvatures depend, up to a factor ± 1, on the choice of normal while Gaussian curvature has the same value for *any* choice of normal. In practice this means that any parametrization may be used to calculate Gaussian curvature while only parametrizations consistent with the choice of normal may be used to find principal curvatures. Indeed, along these lines, we have a celebrated theorem of Gauss – *theorema egregium* – which asserts that Gaussian curvature is an *intrinsic* property of the surface. Roughly speaking this says that Gaussian curvature *may be calculated* directly from functions defined internally on the surface and without using such external properties as the normal or the fact that the surface lies in $\mathbb{R}^3$. Our method of calculating K uses the normal so Gauss' theorem tells us that there is another way of calculating K which does not use the normal. At first glance this may appear a rather minor point but it was this result which paved the way for the development of a very powerful and a very general type of geometry – *Riemannian geometry* – in which the key concepts are *differentiation* and the *length of tangent vectors*. Thus for simplicity, for practical and intrinsic reasons, Gaussian curvature has many advantages. However, in studying surfaces *all* of the concepts we discussed play a useful role and none should be neglected.

So far we have studied the shape of a surface by examining curves (normal sections) of the surface but there are other intuitive approaches to the same problem. Almost invariably they lead back to Gaussian curvature. For example, following the successful approach to plane curvature obtained by taking limits of circles, it is natural to regard the reciprocal of the radius of the sphere that sits closest to the surface near p as a measure of the curvature at p. For surfaces the definition of the *sphere of closest fit* is not so obvious, especially at saddle points, and an indirect approach is taken. We use a normal $\mathbf{n}$ to project the surface near p onto the unit sphere in $\mathbb{R}^3$ and compare the area of the surface near p with the area of its image. When used in this way $\mathbf{n}$ is called the *Gauss map*. If B_ε is the ball with centre p and radius ε then

$$|K(p)| = \lim_{\varepsilon \to 0} \left| \frac{\text{Area}\left(\mathbf{n}(S \cap B_\varepsilon)\right)}{\text{Area}\left(S \cap B_\varepsilon\right)} \right|.$$

Note that if S is a plane then $\mathbf{n}(S)$ consists of a single point and $K(p) = 0$ while if S is a sphere of radius r then $\mathbf{n}(x) = x/r$ for all $x \in S$ and $|K(p)| = 1/r^2$. Further geometric interpretations appear in the final chapter.

We list now a number of results on Gaussian curvature which give some idea of its uses; more advanced and deeper results are given later. Think about these results, ask yourself if they are geometrically plausible, how they fit in with your intuition, what they say about the surfaces with which you are already familiar, and how you might go about proving them.

(i) If S is a connected surface in $\mathbb{R}^3$ consisting entirely of umbilics then S is either an open subset of a sphere or a plane.

(ii) Every compact surface in $\mathbb{R}^3$ contains a point p with $K(p) > 0$.

(iii) A compact connected surface of *constant Gaussian curvature* is a sphere.

(iv) *Hilbert's Lemma:* If p is a non-umbilic point in S, k_1 has a local maximum at p and k_2 has a local minimum at p, then $K(p) \le 0$.

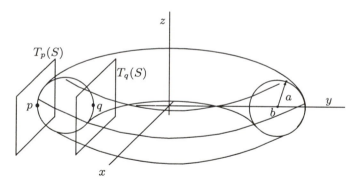

Figure 17.1.

Example 17.1

In this example we discuss the torus. From Figure 17.1 we see that at the point p the surface lies on one side of the tangent plane and $K(p) > 0$ while at q it lies on both sides and $K(q) < 0$.

We use toroidal polar coordinates for our parametrization,

$$P\colon (\theta,\ \phi) \to \big((b + a\cos\theta)\cos\phi,\ (b + a\cos\theta)\sin\phi,\ a\sin\theta\big)$$

$(\theta,\ \phi) \in (0,\ 2\pi) \times (0,\ 2\pi)$. From Examples 11.1 and 14.4, $E = a^2$, $F = 0$, $G = (b + a\cos\theta)^2$, and

$$
\begin{aligned}
\mathbf{n} &= -(\cos\theta\cos\phi,\ \cos\theta\sin\phi,\ \sin\theta) \\
P_\theta &= (-a\sin\theta\cos\phi,\ -a\sin\theta\sin\phi,\ a\cos\theta) \\
P_\phi &= \big(-(b + a\cos\theta)\sin\phi,\ (b + a\cos\theta)\cos\phi,\ 0\big) \\
P_{\theta\theta} &= (-a\cos\theta\cos\phi,\ -a\cos\theta\sin\phi,\ -a\sin\theta) \\
P_{\theta\phi} &= (-a\sin\theta\sin\phi,\ -a\sin\theta\cos\phi,\ 0) \\
P_{\phi\phi} &= \big(-(b + a\cos\theta)\cos\phi,\ -(b + a\cos\theta)\sin\phi,\ 0\big).
\end{aligned}
$$

Hence

$$
\begin{aligned}
l &= \langle P_{\theta\theta}, \mathbf{n}\rangle = a\cos^2\theta\cos^2\phi + a\cos^2\theta\sin^2\phi + a\sin^2\theta = a \\
m &= \langle P_{\theta\phi}, \mathbf{n}\rangle = a\cos\theta\sin\theta\cos\phi\sin\phi - a\cos\theta\sin\theta\cos\phi\sin\phi = 0 \\
n &= \langle P_{\phi\phi}, \mathbf{n}\rangle = (b + a\cos\theta)(\cos\theta\cos^2\phi + \cos\theta\sin^2\phi) \\
&= (b + a\cos\theta)\cos\theta
\end{aligned}
$$

and

$$K = \frac{ln - m^2}{EG - F^2} = \frac{a(b + a\cos\theta)\cos\theta}{a^2(b + a\cos\theta)^2} = \frac{\cos\theta}{a(b + a\cos\theta)}.$$

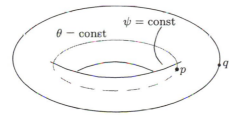

$\psi = \text{const}$

$\theta - \text{const}$

p q

Figure 17.2.

We now have a formula for Gaussian curvature and also a diagram (Figure 17.2) and we may compare and combine them. By differentiating $K(\theta,\ \phi)$, which only depends on θ, we can locate the points of maximum and minimum Gaussian curvature, p and q, and this confirms what the sketch tells us.

Although, in general, it may be difficult to find principal curvatures and principal curvature directions it is possible in special cases and this is the content of the following proposition. We say that a curve Γ on a surface is a *line of curvature* if its tangents are principal curvature directions at each point.

Proposition 17.1

If $\phi: U \subset \mathbb{R}^2 \to S$ is a parametrised surface then the coordinate curves of ϕ are lines of curvature if and only if $F = m = 0$ at all non-umbilic points. In this case the principal curvatures are l/E and n/G. At non-umbilic points the Weingarten mapping L has matrix

$$\begin{pmatrix} l/E & 0 \\ 0 & n/G \end{pmatrix}$$

relative to the orthonormal basis $\left(\dfrac{\phi_x}{\sqrt{E}}, \dfrac{\phi_y}{\sqrt{G}} \right)$ and at umbilic points $L = \dfrac{l}{E} I$ where I is the identity matrix.

Proof

For a parametrised surface we always choose $\phi_x \times \phi_y \, / \, \|\phi_x \times \phi_y\|$ as unit normal. This fixes our choice of shape operator and principal curvatures. We write L, ϕ_x, ϕ_y in place of L_p, $\phi_x(x,y)$, $\phi_y(x,y)$. Since (ϕ_x, ϕ_y) is a basis for the tangent space there exist real-valued functions a, b, c and d on S such that

$$\begin{aligned} L\phi_x &= a\phi_x + b\phi_y \\ L\phi_y &= c\phi_x + d\phi_y. \end{aligned}$$

The coordinate curves $x \to \phi(x,y)$ and $y \to \phi(x,y)$ have tangents $\phi_x(x,y)$ and $\phi_y(x,y)$ respectively and hence will be lines of curvature if and only if $b = c = 0$ at non-umbilic points (at umbilic points all directions are principal curvature directions). We have

$$l = L\phi_x \cdot \phi_x = a\phi_x \cdot \phi_x + b\phi_y \cdot \phi_x = aE + bF \qquad (17.1)$$

$$m = L\phi_x \cdot \phi_y = a\phi_x \cdot \phi_y + b\phi_y \cdot \phi_y = aF + bG \qquad (17.2)$$

$$m = L\phi_y \cdot \phi_x = c\phi_x \cdot \phi_x + d\phi_y \cdot \phi_x = cE + dF \qquad (17.3)$$

$$n = L\phi_y \cdot \phi_y = c\phi_x \cdot \phi_y + d\phi_y \cdot \phi_y = cF + dG. \qquad (17.4)$$

Since E and G are always non-zero, (17.2), (17.3) and $m = F = 0$ imply $b = c = 0$.

Conversely, if $b = c = 0$, then Equations (17.2) and (17.3) imply $m = aF = dF$, $L\phi_x = a\phi_x$ and $L\phi_y = d\phi_y$. Hence $F = 0$ or $a = d$. If $a = d$ at $p = \phi(x,y)$ then p is umbilic. Hence, at non-umbilic points, $a \neq d$ and $F = m = 0$. When $b = c = 0$ we have $a = l/E$ and $d = n/G$. This implies $L\phi_x = (l/E)\phi_x$ and

$L\phi_y = (n/G)\phi_y$ and the principal curvatures are l/E and n/G. It is now easily verified that the Weingarten mapping has the required form.

For the torus $F = m = 0$. Hence $(-a\sin\theta\cos\phi, -a\sin\theta\sin\psi, a\cos\theta)$ and $(-(b+a\cos\theta)\sin\psi, (b+a\cos\theta)\cos\psi, 0)$ are principal curvature directions and the principal curvatures, associated with toroidal polar coordinates, are $1/a$ and $\cos\theta/(b + a\cos\theta)$.

Example 17.2

The surface obtained by rotating the graph of $h\colon (a, b) \to \mathbb{R}$ (Example 10.2) is parametrised by
$$P(t,\, \theta) = (t,\, h(t)\cos\theta,\, h(t)\sin\theta).$$

For this parametrization $P_t = (1, h'\cos\theta, h'\sin\theta)$ and $P_\theta = (0, -h\sin\theta, h\cos\theta)$. Hence $E = 1 + (h')^2$, $F = 0$, $G = h^2$,

$$P_{tt} = (0,\, h''\cos\theta,\, h''\sin\theta)$$

$$P_{t\theta} = (0,\, -h'\sin\theta,\, h'\cos\theta)$$

and

$$P_{\theta\theta} = (0,\, -h\cos\theta,\, -h\sin\theta).$$

Moreover,

$$P_t \times P_\theta = (h'h,\, -h\cos\theta,\, -h\sin\theta)$$

and

$$\mathbf{n} = \frac{(h',\, -\cos\theta,\, -\sin\theta)}{\left(1 + (h')^2\right)^{1/2}}.$$

Hence

$$
\begin{aligned}
l &= \langle P_{tt}, \mathbf{n} \rangle = \frac{-h''}{\left(1 + (h')^2\right)^{1/2}} \\
m &= \langle P_{t\theta}, \mathbf{n} \rangle = 0 \\
n &= \langle P_{\theta\theta}, \mathbf{n} \rangle = \frac{h}{\left(1 + (h')^2\right)^{1/2}}.
\end{aligned}
$$

By Proposition 17.1 the coordinate curves are lines of curvature and the principal curvatures are

$$\frac{l}{E} = \frac{-h''}{\left(1 + (h')^2\right)^{3/2}}, \quad \frac{n}{G} = \frac{1}{h\left(1 + (h')^2\right)^{1/2}} \quad \text{and} \quad K = \frac{-h''}{h\left(1 + (h')^2\right)^2}.$$

An interesting case occurs when $h(t) = c \cosh(t/c)$. This is the shape assumed by a *hanging chain* and is called a *catenary*. The surface of revolution is called a *catenoid*. Using the identities

$$\frac{d}{dt}(\cosh t) = \sinh t, \quad \frac{d}{dt}(\sinh t) = \cosh t, \quad \cosh^2 t - \sinh^2 t = 1,$$

we see that the principal curvatures for the catenoid are $\pm 1/c \cosh^2(t/c)$. Hence $k_1 + k_2 = 0$. A surface with this property is called a *minimal surface*. This terminology arose in the following way. Take a closed curve in $\mathbb{R}^3$ – shaped, for instance, from a piece of wire – and place a bubble over it. This will assume a certain shape in order to *minimise* a physical quantity on the boundary called surface tension. The shape assumed by the bubble is a minimal surface. The catenoid is the only surface of revolution which is a minimal surface. On minimal surfaces $k_2(p) = -k_1(p)$ and hence

$$K(p) = k_1(p)k_2(p) = -k_1(p)^2 \leq 0.$$

The quantity $\dfrac{k_1(p) + k_2(p)}{2}$ is called the *mean curvature*.

EXERCISES

17.1 Find the Weingarten mapping for the torus.

17.2 Let **S** denote the graph of the function $f(u, v) = u^2 - 2v^2$ oriented by the usual parametrization (Example 10.1). By using the result in Exercise 16.4 or otherwise find a curve Γ in **S** such that the normal curvature in the tangent direction to Γ is always zero.

17.3 Suppose the two oriented surfaces S_1 and S_2 intersect in a curve Γ. Let κ denote the curvature of $\Gamma \subset \mathbb{R}^3$ and let λ_i denote the normal curvature of Γ in S_i, $i = 1, 2$. If θ is the angle between the normals to S_1 and S_2 show that

$$\kappa^2 \sin^2 \theta = \lambda_1^2 + \lambda_2^2 - 2\lambda_1\lambda_2 \cos \theta.$$

17.4 If $P: U \subset \mathbb{R}^2 \to \mathbb{R}^3$ is a parametrization of a simple surface S show that a point p is an umbilic point if and only if there exists a real number α such that $(l, m, n) = \alpha(E, F, G)$ where each term is evaluated at p. Find all umbilics on the surface $z = xy$.

17.5 Show that the average of the normal curvature over all directions is the mean curvature.

17.6 Show that $(\kappa_1(p) + \kappa_2(p))/2 = 1/2$ (sum of diagonal entries in L_p). Hence, show using the matrix representation given in Chapter 16, that a surface S is a minimal surface if and only if

$$lG + En - 2Fm = 0.$$

Show that the level set $e^z \cos x = \cos y$ is a minimal surface.

17.7 Consider the plane curve h described by the following geometric condition: start at the point $(0, c)$ and move so that the tangent line always reaches the x-axis after travelling a distance c. From Figure 17.3 deduce that

$$h' = -\frac{h}{\sqrt{c^2 - h^2}}.$$

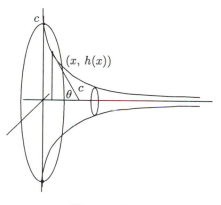

Figure 17.3.

(Note that $\sin\theta = -h/c$, $\tan\theta = h' = \dfrac{\sin\theta}{\sqrt{1 - \sin^2\theta}}$.) Show that

$$h'' = \frac{c^2 h}{(c^2 - h^2)^2}.$$

Orient the surface of revolution, B, of this curve so that the principal curvatures are h'/c and $-1/ch'$. Hence deduce that B has constant negative Gaussian curvature. Sketch the surface B – it may help you to know that it is called the *bugle surface*.

18

Geodesic Curvature

Summary. *We define geodesic curvature and geodesics. For a curve on a surface we derive a formula connecting intrinsic curvature, normal curvature and geodesic curvature. We discuss paths of shortest distance, further interpretations of Gaussian curvature and introduce, informally and geometrically, a number of important results in differential geometry.*

Our study of normal curvature was based on identifying the normal of a curve with the normal of the surface. It is also possible to arrange things so that the normal to a curve in a surface is a tangent vector to the surface. This leads to a new type of curvature, *geodesic curvature*, that we discuss and interpret in this chapter.

Let $\mathbf{S}$ denote an oriented surface in $\mathbb{R}^3$ with smooth unit normal $\mathbf{n}$ and let $P:[a, b] \to \Gamma \subset \mathbf{S}$ denote a unit speed parametrised curve on the surface. Since the tangent space at each point on the surface is two-dimensional and $P'(t)$ is a tangent vector at $P(t)$ it follows that there are precisely two unit tangent vectors at $P(t)$ which are perpendicular to $P'(t)$. We distinguish between them by using the normal to the surface at $P(t)$, $\mathbf{n}(P(t))$. To simplify our notation we write $\mathbf{n}(t)$ in place of $\mathbf{n}(P(t))$ from now on.

We define the *surface normal* to Γ at $P(t)$, $\mathbf{n_S}(t)$ to be $\mathbf{n}(t) \times T(t)$. It follows immediately that $\{P'(t) = T(t), \mathbf{n}_S(t), \mathbf{n}(t)\}$ is a right-handed orthogonal system, and in particular, $\{T(t), \mathbf{n_S}(t), \mathbf{n}(t)\}$ is an orthonormal basis for $\mathbb{R}^3$. In Chapter 7 we encountered a similar situation when we obtained the orthonormal basis $\{T, N, B\}$ at a point on a curve Γ. In that case we proceeded to obtain the Frenet–Serret equations by differentiation and using properties of

223

the orthonormal basis. We follow *precisely* the same path to a similar end here and obtain real-valued functions $a(t)$, $b(t)$ and $c(t)$ such that

$$
\begin{pmatrix} T(t) \\ \mathbf{n_S}(t) \\ \mathbf{n}(t) \end{pmatrix}' = \begin{pmatrix} 0 & a(t) & c(t) \\ -a(t) & 0 & b(t) \\ -c(t) & -b(t) & 0 \end{pmatrix} \begin{pmatrix} T(t) \\ \mathbf{n_S}(t) \\ \mathbf{n}(t) \end{pmatrix}.
$$

In particular we see that

$$
\mathbf{n_S}'(t) = -a(t)T(t) + b(t)\mathbf{n}(t) \tag{18.1}
$$

and the "part" of $\mathbf{n_S}'(t)$ which "lies" in the tangent space at $P(t)$ is parallel to $T(t)$. This equation is similar to Equation (7.9) in Chapter 7. We define the *geodesic curvature* of Γ at $P(t)$, $\kappa_g(t)$, to be $a(t)$. By (18.1)

$$
\kappa_g(t) = -\langle \mathbf{n_S}'(t), T(t) \rangle = a(t). \tag{18.2}
$$

The entry $b(t)$ is called the *geodesic torsion* of Γ at $P(t)$ and written $\tau_g(t)$. Rewriting Equation (18.1) we obtain

$$
\mathbf{n_S}'(t) = -\kappa_g T(t) + \tau_g \mathbf{n}(t).
$$

If $\mathbf{v}$ is a unit tangent vector at $p \in \mathbf{S}$ then

$$
\langle L_p(\mathbf{v}), \mathbf{v} \rangle = \kappa_p(\mathbf{v})
$$

and

$$
L_p(\mathbf{v}) = -\frac{d}{dt}\Big(\mathbf{n}\big(\alpha(t)\big)\Big)\Big|_{t=0}
$$

for any curve α satisfying $\alpha(0) = p$ and $\alpha'(0) = \mathbf{v}$.

Hence

$$
-\langle \mathbf{n}'(t), T(t) \rangle = -\langle \frac{d}{dt}\mathbf{n}(t), T(t) \rangle = \kappa_{P(t)}\big(T(t)\big) = c(t)
$$

i.e. $c(t)$ is the *normal curvature* at $P(t)$ in the direction $T(t)$, and

$$
\begin{pmatrix} T(t) \\ \mathbf{n_S}(t) \\ \mathbf{n}(t) \end{pmatrix}' = \begin{pmatrix} 0 & \kappa_g(t) & \kappa_n(t) \\ -\kappa_g(t) & 0 & \tau_g(t) \\ -\kappa_n(t) & -\tau_g(t) & 0 \end{pmatrix} \begin{pmatrix} T(t) \\ \mathbf{n_S}(t) \\ \mathbf{n}(t) \end{pmatrix}
$$

where we have written $\kappa_n(t)$ in place of $\kappa_{P(t)}\big(T(t)\big)$. This implies

$$
T'(t) = \kappa_g(t)\mathbf{n_S}(t) + \kappa_n(t)\mathbf{n}(t).
$$

If we forget about $\mathbf{S}$ and think of Γ as a parametrised curve in $\mathbb{R}^3$ then we can calculate the curvature κ discussed in Chapter 7 – to avoid confusion between the different types of curvature we now call κ the *intrinsic curvature* of Γ.

From the first of the Frenet–Serret equations for a curve in $\mathbb{R}^3$ (Chapter 7) we have

$$T'(t) = \kappa(t)N(t)$$

where $N(t)$ is the normal of Γ in $\mathbb{R}^3$ at $P(t)$. Comparing these expressions for $T'(t)$ we obtain

$$\kappa(t)N(t) = \kappa_g(t)\mathbf{n_S}(t) + \kappa_n(t)\mathbf{n}(t) \tag{18.3}$$

and, by Pythagoras' theorem,

$$\kappa^2(t) = \kappa_g^2(t) + \kappa_n^2(t)\,.$$

Equation (18.3) is a decomposition of the intrinsic curvature into its normal and tangential components and establishes a relationship between the three different kinds of curvature. If we consider curvature as a measure of "bending" towards the normal then, since we have chosen our normal $\mathbf{n_S}(t)$ to lie in the tangent space, this and (18.3) suggest that we consider geodesic curvature as the surface curvature of the parametrised curve. So far we have only considered a unit speed parametrised curve. We define the geodesic curvature of a parametrised curve in an oriented surface as the geodesic curvature of a unit speed reparametrization of the curve which preserves the original sense of direction. Note that our definition of geodesic curvature is based, as was normal curvature, on curvature in $\mathbb{R}^2$ – see the introduction to Chapter 7. Since the two unit normals at any point on a surface are parallel, Equation (18.3) shows that $|\kappa_g|$ – the *absolute geodesic curvature* – does not depend on either the choice of normal or parametrization. As surfaces can be covered by simple, and hence orientable, surfaces it follows that absolute geodesic curvature is *well defined* for *any* curve on *any* surface.

An important class of curves are those with $|\kappa_g| = 0$. Curves with this property are said to have *zero geodesic curvature*. We give a geometrical interpretation of this phenomena and afterwards discuss a practical method for identifying such curves. If we are dealing with a curve in $\mathbb{R}^2$ then zero curvature implies that the curve is a straight line. If we identify $\mathbb{R}^2$ with the oriented surface $\mathbb{R}^2_{(x,y)}$ in $\mathbb{R}^3$ and consider a curve Γ in $\mathbb{R}^2$ as a curve in $\mathbb{R}^3$ then we see easily that its plane curvature in $\mathbb{R}^2$ coincides with its geodesic curvature in $\mathbb{R}^2_{(x,y)}$. In $\mathbb{R}^2$ we also note that a curve is a straight line if and only if it follows the shortest route between any pair of its points. Now the *tangent plane* is the *closest plane* to the surface near a given point p and since geodesic curvature is essentially curvature on the tangent plane it is at least plausible that zero geodesic curvature implies that Γ follows the *shortest path on the surface* between points on Γ close to p. This is indeed the case.

Formally we have the following definitions and results. A surface S is *connected* if between any two points p and q on S there exists at least one path (or directed curve) with initial point p and final point q. We define the *distance* between p and q, $d(p, q)$, as

$$\inf \{\text{length}(\gamma), \gamma \text{ is a path with initial point } p \text{ and final point } q\}.$$

A path γ joining p and q is a shortest path if

$$d(p, q) = \text{length}(\gamma).$$

Shortest paths may or may not exist and if they exist they may not be unique. For instance it is easily seen that there is no shortest path on the surface $S = \{(x, y, 0) : 0 < x^2 + y^2 < 2\}$ in $\mathbb{R}^3$ joining the points $(-1, 0, 0)$ and $(1, 0, 0)$ although it is easy to see that the distance between them is 2. On the other hand there exist an infinite number of shortest paths on a sphere joining the North and South poles (any line of longitude is a shortest path). Equation (18.3) leads to a simple practical criterion for identifying unit speed curves of zero geodesic curvature since it is easily seen that $P''(t)$ (or $N(t)$) is parallel to $\mathbf{n}(t)$, for any choice of normal, if and only if $|\kappa_g(t)| = 0$. We can restate this observation so that the normal is not mentioned and as a result we do not require an *oriented surface*. This is the basis for the following definition.

Definition 18.1

A parametrised curve $P : [a, b] \to \Gamma \subset S$, where S is a surface in $\mathbb{R}^3$, is a geodesic in S if $P''(t)$ (the acceleration of the parametrization) is perpendicular to the tangent space at $P(t)$) for all t.

Our remarks above show that *a unit speed parametrised curve is a geodesic if and only if it has zero geodesic curvature.*

Lemma 18.1

Geodesics have constant speed.

Proof

We have $\dfrac{d}{dt} \langle P'(t), P'(t) \rangle = 2 \langle P''(t), P'(t) \rangle$. Since the curve lies in S, $P'(t)$ is a tangent vector at $P(t)$, and as P is a geodesic $P''(t)$ is perpendicular to every tangent vector. This implies $\langle P'(t), P''(t) \rangle = 0$ and $\dfrac{d}{dt} \left(\|P'(t)\|^2 \right) = 0$. Hence $\|P'(t)\|$ is a constant function of t. This completes the proof.

Now suppose the curve parametrised by P has constant non-zero speed c. We may suppose, without loss of generality, that P is defined on the interval $[0, l]$. The parametrised curve defined by $Q(t) = P(t/c)$ on $[0, lc]$ has unit speed and $Q''(t) = (1/c^2)P''(t/c)$. Since $P''(t/c)$ is parallel to $Q''(t)$ it follows that P is a geodesic if and only if the unit speed curve Q is a geodesic. We have already noted that Q is a geodesic if and only if it has zero geodesic curvature and since P and Q have, by definition, the *same* geodesic curvature we have proved the following proposition.

Proposition 18.1

A parametrised curve $P \colon [a, b] \to S$ is a geodesic if and only if it has constant speed and zero geodesic curvature.

From the above considerations it is not difficult to show that a parametrised curve is a geodesic if and only if it satisfies a certain ordinary differential equation. Results from the theory of ordinary differential equations show that any surface admits an abundance of geodesics. The following is true.

Proposition 18.2

If S is a surface in $\mathbb{R}^3$, $p \in S$ and $\mathbf{v} \in T_p(S)$ is non-zero then there exists for ε sufficiently close to zero a unique geodesic $P \colon [-\varepsilon, \varepsilon] \to S$ such that $P(0) = p$ and $P'(0) = \mathbf{v}$.

Example 18.1

Let S_2 denote the unit sphere with centre at the origin in $\mathbb{R}^3$. Let $\mathbf{v}$ and $\mathbf{w}$ denote perpendicular unit vectors in $\mathbb{R}^3$ and let

$$P(t) = \cos(at)\mathbf{v} + \sin(at)\mathbf{w}, \quad t \in \mathbb{R}$$

where a is a fixed real number. By Pythagoras' theorem P defines a parametrised curve in S_2. We have

$$P''(t) = -a^2 \cos(at)\mathbf{v} - a^2 \sin(at)\mathbf{w} = -a^2 P(t).$$

Since $\mathbf{n}(t) = \pm P(t)$ for the unit sphere with centre at the origin this implies $P''(t) \parallel \mathbf{n}(t)$ for all t and P is a geodesic. Proposition 18.2 shows that a geodesic is completely determined by its position and velocity at a single point. Since $P(0) = \mathbf{v}$ and $P'(0) = a\mathbf{w}$ it follows that we have found *all* geodesics on the unit sphere.

We return to the problem we started with – the existence of a shortest path. Propositions 18.1 and 18.2 show that there are many curves on a surface with zero geodesic curvature. From this it is possible to prove the following result which shows, at least locally, that there are always shortest paths.

Proposition 18.3

If p is a point on a surface S in $\mathbb{R}^3$ then there exists $\varepsilon > 0$ such that for any q in S, $d(p,q) < \varepsilon$, there is a unique shortest path in S joining p and q. This path has zero geodesic curvature and may be parametrised as a geodesic.

Geodesics also lead to a new derivation of Gaussian curvature. Take a point p on the surface S. An extension of Proposition 18.2 shows that there exists $\varepsilon > 0$ such that for every unit tangent vector $\mathbf{v}$ at p the geodesic with initial point p and initial velocity $\mathbf{v}$ is defined on $[0, \varepsilon]$. If we consider the set of positions taken at time ε by all unit speed geodesics starting at p we obtain a curve γ_ε in S. We denote the inside of this curve by A_ε (Figure 18.1).

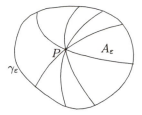

Figure 18.1.

Thus γ_ε consists of those points in S whose distance to p is ε and A_ε are the points whose distance to p is less than ε. It can be shown that γ_ε is a closed subset of S and A_ε is open. If S is flat then γ_ε is a circle of radius ε and length $2\pi\varepsilon$ and A_ε is a disc of area $\pi\varepsilon^2$. Hence the quantities $2\pi\varepsilon - l(\gamma_\varepsilon)$ and $\pi\varepsilon^2 - \mathrm{A}(A_\varepsilon)$ where $l = $ length and $\mathrm{A} = $ Area are some measure of the curvature of S. In fact, taking limits we obtain the following

$$K(p) = \lim_{\varepsilon \to 0} \frac{3}{\pi\varepsilon^3}\left(2\pi\varepsilon - l(\gamma_\varepsilon)\right) = \lim_{\varepsilon \to 0} \frac{12}{\pi\varepsilon^4}\left(\pi\varepsilon^2 - A(A_\varepsilon)\right).$$

Thus both normal and geodesic curvature lead quite naturally to Gaussian curvature. It is also worth noting that the two most important geometrical concepts that we associated with a surface – Gaussian curvature and geodesics – both turned out to be local properties independent of any orientations used for calculations or motivation.

To complete our introduction to the geometry of surfaces we present without proof some rather remarkable results involving the concepts we have introduced. These results are easily stated while the proofs are rather involved. Even in the absence of any ideas regarding the proofs it is well worth thinking about these results and their geometric significance. You may consider verifying them on some of the classical surfaces we have studied, e.g. the sphere, ellipsoid or torus. Trying to prove them will take some time but if you make the effort and are patient you will learn a lot regardless of how successful you are in completing the proofs.

The first result we discuss is a generalization of the well-known result in Euclidean geometry which says that the sum of the interior angles in a (plane) triangle is equal to π. This corresponds to the case where the surface is a plane and the Gaussian curvature is zero. A *triangle* in a surface is a simple closed oriented curve formed by three smooth directed curves (Figure 18.2).

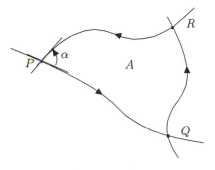

Figure 18.2.

Each of these smooth curves is called an *edge*, the edges meet at a *vertex* and the interior of the triangle is called its *face*. A triangle is called a *geodesic triangle* if each edge is a geodesic. In the plane, geodesics are straight lines so the usual triangle in the plane is a geodesic triangle. At the vertices P, Q and R the tangents of the two curves which meet are in the same tangent space on the surface and we can define the *angle* between them. It is also possible to define what we mean by an *interior angle* (e.g. the angle α at the vertex P). Let A denote the face of the triangle.

Theorem 18.1 (Local Gauss–Bonnet Theorem)

In a geodesic triangle on a simple surface

$$\iint\limits_A K = \sum \text{interior angles} - \pi.$$

Stokes' theorem plays an important role in the proof. For example consider the sphere of radius r. By Example 18.1 the lines of longitude and the equator are geodesics. Hence the triangle formed by the lines of longitude corresponding to $\psi = 0$ and $\psi = \pi/2$ and the equator are a geodesic triangle (Figure 18.3).

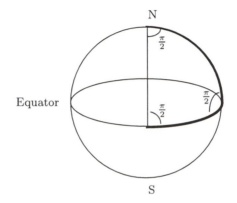

Figure 18.3.

The three interior angles are all $\pi/2$ and the area of the triangle is $\frac{1}{8} \times$ (total area of the sphere) and hence equal to $4\pi r^2/8$. Since the Gaussian curvature is $1/r^2$ this implies

$$\iint\limits_A K = \frac{1}{r^2} \cdot \frac{4\pi r^2}{8} = \frac{\pi}{2}$$

while

$$\sum \text{interior angles} - \pi = \frac{3\pi}{2} - \pi = \frac{\pi}{2}$$

and we have verified the local Gauss–Bonnet theorem for this triangle.

Any compact (i.e. closed and bounded) oriented surface $\mathbf{S}$ in $\mathbb{R}^3$ can be partitioned into a finite number of triangles each of which is oriented in an anticlockwise direction about the normal (Figure 18.4(a)). This means that an edge which is in two triangles has opposite orientations in each (Figure 18.4(b)). Let V denote the total number of vertices, E the total number of edges and F the total number of faces.

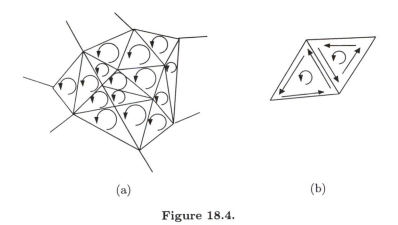

(a)   (b)

Figure 18.4.

A remarkable result of Euler says that no matter how we partition the surface into (not necessarily geodesic) triangles the quantity $V - E + F$ remains unchanged. We call this number the *Euler–Poincaré characteristic* of S and denote it by $\chi(S)$. As a consequence of the local Gauss–Bonnet theorem we have the following global result:

Theorem 18.2 (Global Gauss–Bonnet Formula)

If S is a compact oriented surface then

$$\iint_S K = 2\pi\chi(S).$$

We call $\iint_S K$ the *total curvature* of the surface and it is remarkable that this quantity is always an integer multiple of 2π. Applying this to a sphere of radius r we find

$$\iint_S K = \frac{1}{r^2} \text{ Surface Area (sphere)} = \frac{4\pi r^2}{r^2} = 2\pi\chi(S)$$

and see that $\chi(\text{sphere}) = 2$. On the other hand we may partition the sphere by the lines of longitude corresponding to 0, $\pm\pi/2$, π and the equator (Figure 18.5).

The outer edges are on the back of the sphere and coincide with $180°$ East or $180°$ West (i.e. the international date line). By counting we get $V = 6$, $E = 12$ and $F = 8$ and again

$$\chi \text{ (sphere)} = 6 - 12 + 8 = 2\,.$$

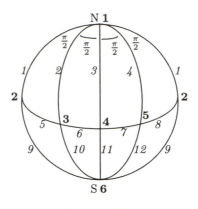

Figure 18.5.

Any compact oriented surface can be smoothly deformed into a surface with a finite number of holes (Figure 18.6).

We call the number of holes the *genus* of the surface. This smooth deformation does not change the number of faces, edges or vertices of any partition and hence the Euler–Poincaré characteristic is unchanged. It does, however, change the Gaussian curvature in many places, e.g. the sphere can be changed into an ellipsoid and we know that these do not have the same Gaussian curvature.

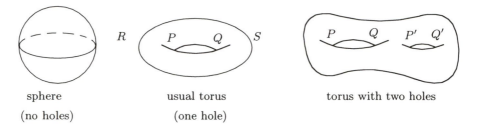

sphere   usual torus   torus with two holes

(no holes)   (one hole)

Figure 18.6.

The global Gauss–Bonnet theorem says that the total curvature is unchanged. We also have

$$\chi(S) = 2 - 2g$$

where g is the genus. This implies that the Euler–Poincaré characteristic is always an even integer and the total curvature is an integer multiple of 4π.

Thus we have a remarkable set of relationships between total curvature, the number $V - E + F$ and the number of holes on a compact oriented surface **S**. This is not the end of the story as they are all equal to the index of any smooth vector field on **S**. The *index* of a vector field X, $i(X)$, is obtained by assigning an integer to each zero following a prescribed formula. Since χ (sphere) $= 2$ it follows that every smooth tangent vector field on a sphere has at least one zero and explains the existence of the bald spot which most people usually have at the point of maximal curvature on the midline of the calva.

It is interesting to speculate in a purely geometric way why these things are the way they are. For instance, we have seen that the total curvature of a sphere does not depend on the radius. Think of blowing up a balloon. The bigger the radius the larger the surface area. On the other hand the sphere is becoming less curved, i.e. the Gaussian curvature is decreasing, and over the whole surface the increase in surface area is counterbalanced by the decrease in curvature. Another simple observation: from the formula $\chi(S) = 2 - 2g$ we see that adding holes decreases the Euler–Poincaré characteristic and hence adds *negative* Gaussian curvature to the surface. Why should this be so? If we recall our study of the Gaussian curvature of the torus (Example 17.1) we noted that at the points P and Q in Figure 18.6 we had negative Gaussian curvature while at R and S we had positive Gaussian curvature. Adding holes creates points like P' and Q' while the outside, where the Gaussian curvature is positive, is relatively unchanged and it is at least plausible that we are increasing the overall negative Gaussian curvature by adding holes.

EXERCISES

18.1 Show that every geodesic on the cylinder $\{(x, y, z) : x^2 + y^2 = 1\}$ has the form

$$\phi(t) = \big(\cos(at + b), \sin(at + b), ct\big), \quad t \in \mathbb{R}.$$

18.2 Let $\phi(t) = \big(x(t), y(t)\big)$, $a < t < b$ denote a unit speed parametrised curve in $\mathbb{R}^2$ and suppose $y(t) > 0$. Let S denote the surface obtained by rotating this curve about the x-axis. If

$$P(t, \theta) = \big(x(t), y(t) \cos \theta, y(t) \sin \theta\big)$$

show that the mapping

$$t \longrightarrow P(t, \theta_0)$$

is a geodesic for every fixed θ_0.

18.3 Prove that a straight line which lies in a surface has zero geodesic curvature on the surface. By using this result find for each point P on the surface $z = x^2 - y^2$ two geodesics passing through P.

18.4 Let $P \colon [a,b] \to \Gamma$ denote a unit speed parametrised curve on the surface S. Show that Γ is a line of curvature if and only if the geodesic torsion $\tau_g = 0$. If Γ is a geodesic show that $\tau_g = \tau$ where τ is the torsion of Γ as a curve in $\mathbb{R}^3$ (Chapter 7).

18.5 Show that a curve in a sphere with constant geodesic curvature is part of a circle.

18.6 Let $P \colon [a,b] \to \Gamma$ denote a unit speed curve in an oriented surface $\mathbf{S}$. Show that the normal curvature at $P(t)$ in the direction $P'(t)$ is zero if and only if the osculating plane to the curve coincides with the tangent plane to the surface. (A curve with this property at all points is called an *asymptotic curve* on the surface.)

18.7 Find the Euler–Poincaré characteristic of the torus by partitioning it into triangles and calculating $V - E + F$. Verify your result by calculating the total curvature (see Example 17.1).

Solutions

Solutions, answers, hints or relevant remarks to selected exercises are provided. The unexplained notation can be found in the text.

Chapter 1

1.1 The only non-empty open and closed subset of $\mathbb{R}^n$ is $\mathbb{R}^n$. The union of open sets is open and the intersection of closed sets is closed.

    (a) Interior of a solid ellipse, open and bounded, not compact.

    (b) Surface of a cone, closed, not bounded since $(n, 0, n)$ lies in the surface for all n, not compact.

    (c) First octant in $\mathbb{R}^3$ – like the first quadrant in $\mathbb{R}^2$. Closed, not bounded since $(n, 0, 0)$ lies in the set for all n. Hence not compact.

    (d) $x^2 + y^2 + (z - 1)^2 = 1$ is the surface of a sphere with centre $(0, 0, 1)$ and radius 1. Closed, bounded and hence compact.

    (e) Intersection of sphere of radius 2 and centre $(0, 0, 2)$ and cylinder parallel to z-axis based on circle in (x, y)-plane with centre $(0, 0)$ and radius 2. Intersection is circle in the plane $z = 2$ of radius 2 with centre $(0, 0, 2)$, compact.

1.2 (a) $\frac{\partial f}{\partial x} = 2x \log(1 + x^2 y^2) + \frac{2(z^2 + x^2)xy^2}{1 + x^2 y^2}$, $\frac{\partial f}{\partial y} = \frac{2(z^2 + x^2)x^2 y}{1 + x^2 y^2}$, $\frac{\partial f}{\partial z} = 2z \log(1 + x^2 y^2)$.

    (b) $\frac{\partial g}{\partial x} = y \tan^{-1}(xz) + \frac{xyz}{1 + x^2 z^2}$, $\frac{\partial g}{\partial y} = x \tan^{-1}(xz)$, $\frac{\partial g}{\partial z} = \frac{x^2 y}{1 + x^2 z^2}$.

1.3 $F'(x, y, z, w) = \begin{pmatrix} 2x & -2y & 0 & 0 \\ 2y & 2x & 0 & 0 \\ z & 0 & x & 0 \\ 2xz^2w^2 & 0 & 2zw^2x^2 & 2z^2wx^2 \end{pmatrix}$,

$D_\mathbf{v}F(1, 2, -1, -2) = \begin{pmatrix} 2 & -4 & 0 & 0 \\ 4 & 2 & 0 & 0 \\ -1 & 0 & 1 & 0 \\ 8 & 0 & -8 & -4 \end{pmatrix} \begin{pmatrix} 2 \\ 1 \\ -2 \\ -1 \end{pmatrix} = \begin{pmatrix} 0 \\ 10 \\ -4 \\ 36 \end{pmatrix}$.

1.4 $\nabla f(x, y, z) = (2x - y, -x + z^3, 3y^2 - 6)$. $\nabla f(x, y, z) = (0, 0, 0) \Leftrightarrow 2x - y, -x + z^3 = 0, 3yz^2 - 6 = 0 \Leftrightarrow y = 2x, z^3 = x, yz^2 = 2 \Leftrightarrow y = 2x, z^3 = x, xz^2 = 1 \Leftrightarrow y = 2x, z^3 = x, z^5 = 1 \Leftrightarrow z = 1, x = 1, y = 2$. Solution $(1, 2, 1)$.

1.5 $\nabla f = (2xe^y, x^2e^y, 0)$, $\nabla g = (zy^2e^{xz}, 2ye^{xz}, xy^2e^{xz})$, $fg = x^2y^2e^{y+xz}$. $\nabla(fg) = (2xy^2e^{y+xz} + x^2y^2ze^{y+xz}, 2x^2ye^{y+xz} + x^2y^2e^{y+xz}, x^3y^2e^{y+xz}) = x^2e^y(zy^2e^{xz}, 2ye^{xz}, xy^2e^{xz}) + y^2e^{xz}(2xe^y, x^2e^y, 0) = f\nabla g + g\nabla f$.

1.6 If $P(t) = (x_1(t), x_2(t), \ldots, x_n(t))$ then $\|P(t)\|^2 = x_1^2(t) + \cdots + x_n^2(t)$ and $\frac{d}{dt}(\|P(t)\|^2) = 2x_1(t)x_1'(t) + \cdots + 2x_n(t)x_n'(t) = 2\sum_{i=1}^n x_i(t)x_i'(t) = 2P(t) \cdot P'(t)$. If $\|P(t)\|$ does not depend on t then $\frac{d}{dt}(\|P(t)\|^2) = 0$ and $P(t) \cdot P'(t) = 0$. The inner product of two vectors is zero if and only if they are perpendicular. In this exercise, which is extensively used in Chaps. 5–8 and 16–18, we used the inner product notation. If we use matrix notation then $P(t)$ is a $1 \times n$ matrix and $P'(t)$ is an $n \times 1$ matrix.

1.7 $F'(X) = \begin{pmatrix} 2x & 0 & 0 \\ 0 & 2y & 2z \\ yz & xz & xy \end{pmatrix}$, $G'(X) = \begin{pmatrix} e^x & 0 & 0 \\ 0 & 2y & -2z \\ yz & xz & xy \end{pmatrix}$

$H(x, y, z) = x^2e^x + y^4 - z^4 + x^2y^2z^2$,

$\nabla H(x, y, z) = (2xe^x + x^2e^x + 2xy^2z^2, 4y^3 + 2x^2yz^2, -4z^3 + 2x^2y^2z)$.

Note: $F: \mathbb{R}^3 \to \mathbb{R}^3$ hence F' is a 3×3 matrix, $G: \mathbb{R}^3 \to \mathbb{R}^3$ hence G' is a 3×3 matrix and $\langle F, G\rangle: \mathbb{R}^3 \to \mathbb{R}^1$ hence $\nabla(\langle F, G\rangle)$ is a 1×3 matrix. This means $G \circ F' + F \circ G'$ is a 1×3 matrix and $F' \circ G + F \circ G'$ would not give the correct answer since it is not possible to add a 1×3 matrix and a 3×1 matrix.

1.8 $\frac{\partial}{\partial x}\left((x^2 + y^2 + z^2)^{-1/2}\right) = -\frac{1}{2}(x^2 + y^2 + z^2)^{-3/2} \cdot 2x = \frac{-x}{(x^2+y^2+z^2)^{3/2}}$.

1.9 Let $F = (f_1, \ldots, f_m)$. Then $\|F\| = (f_1^2 + \cdots + f_m^2)^{1/2}$ and $\frac{\partial}{\partial x_i}(\|F\|) = \frac{1}{2}(f_1^2 + f_2^2 + \cdots + f_m^2)^{-1/2} \cdot \frac{\partial}{\partial x_i}(\sum_{j=1}^m f_j^2) = \frac{1}{\|F\|} \cdot \sum_{j=1}^m 2f_j \frac{\partial f_j}{\partial x_i}$
$= \langle F, \frac{\partial F}{\partial x_i}\rangle / \|F\|$, $\nabla_\mathbf{v}(\|F\|) = \sum_{i=1}^n v_i \frac{\partial}{\partial x_i}(\|F\|) = \frac{1}{\|F\|}\sum_{i=1}^n v_i\langle F, \frac{\partial F}{\partial x_i}\rangle =$

$\frac{1}{\|F\|}\langle F, \sum_{i=1}^{n} v_i \frac{\partial F}{\partial x_i}\rangle = \frac{\langle F, D_{\mathbf{v}}F\rangle}{\|F\|}$. We require $F(p) \neq 0$ since we divide by $\|F\|$. The result is not true otherwise, e.g. $f(x) = |x|$, $x \in \mathbb{R}$, is not differentiable at the origin. Note the use of the notation $\nabla_{\mathbf{v}}$ in place of $D_{\mathbf{v}}$ for scalar-valued functions.

1.10 Let $F(x_1, \ldots, x_n) = (x_1, \ldots, x_n)$ in Exercise 1.9. Then $D_{\mathbf{e}_i}F = \mathbf{e}_i$ and
$D_{\mathbf{e}_i}(1/\|X\|) = -\frac{1}{\|X\|^2}D_{\mathbf{e}_i}(\|X\|) = -\frac{1}{\|X\|^2}D_{\mathbf{e}_i}(\|F\|) = -\frac{\langle F, D_{\mathbf{e}_i}F\rangle}{\|X\|^2\|F\|} = -\frac{\langle \mathbf{e}_i, F\rangle}{\|X\|^3} = -x_i/\|X\|^3$. Hence $\nabla(1/\|X\|) = -X/\|X\|^3$

1.11 $H(x, y, z) = \big((xyz)^2 + (x^2+y^2)^2, (xyz)^2 - (x^2+y^2)^2, (x^2-y^2)^2 - z^4, (x^2-y^2)^2 + z^4\big)$ and $H_2(x, y, z) = (xyz)^2 - (x^2+y^2)^2$. $\frac{\partial H_2}{\partial x} = 2xy^2z^2 - 4x(x^2+y^2)$.

$$G' \circ F' = \begin{pmatrix} 2u & 2v & 0 & 0 \\ 2u & -2v & 0 & 0 \\ 0 & 0 & 2w & -2t \\ 0 & 0 & 2w & 2t \end{pmatrix} \begin{pmatrix} yz & xz & xy \\ 2x & 2y & 0 \\ 2x & -2y & 0 \\ 0 & 0 & 2z \end{pmatrix} = \begin{pmatrix} - & - & - \\ \frac{\partial H_2}{\partial x} & - & - \\ - & - & - \\ - & - & - \end{pmatrix}$$

$\frac{\partial H_2}{\partial x} = 2uyz - 2v \cdot 2x = 2xyz \cdot yz - 2(x^2+y^2)2x = 2xy^2z^2 - 4x(x^2+y^2)$.

1.12 h is the composition

$$(x_1, \ldots, x_n) \rightarrow (e^{x_1}, \ldots, e^{x_n})$$
$$\|$$
$$(y_1, \ldots, y_n) \rightarrow f(y_1, \ldots, y_n)$$

$\frac{\partial h}{\partial x_i} = e^{x_i}\frac{\partial f}{\partial y_i}$, $\frac{\partial^2 h}{\partial x_i^2} = e^{x_i}\frac{\partial f}{\partial y_i} + e^{x_i}\frac{\partial}{\partial y_i}(\frac{\partial f}{\partial y_i}) = e^{x_i}\frac{\partial f}{\partial y_i} + e^{2x_i}\frac{\partial^2 f}{\partial y_i^2} = y_i\frac{\partial f}{\partial y_i} + y_i^2\frac{\partial^2 f}{\partial y_i^2}$. Hence $\sum_{i=1}^{n}\frac{\partial^2 h}{\partial x_i^2} = \sum_{i=1}^{n} y_i\frac{\partial f}{\partial y_i} + \sum_{i=1}^{n} y_i^2\frac{\partial^2 f}{\partial y_i^2} = 0$.

1.14 $P = (1,1,1)$, $\Delta X = (0.1, 0.05, -0.05)$, $f(P) = 3$, $f(P+\Delta X) = 3.42628125$, $\nabla f(P) = (3,5,3)$, $f(P)+\nabla f(P)\cdot\Delta X = 3.40$, Error= 0.02628125, Error$\times 100/f(P) = 0.87604166\%$.

1.15 $F^{-1}(C)$ is the intersection of the cone $z^2 - x^2 - y^2 = 1$ and the plane $2x - y = 2$. Solving these equations yields $y = 2x - 2$ and $z^2 = 1 + x^2 + 4(x-1)^2 = 1 + 5x^2 - 8x + 4 = 9/4 + 5(x-4/5)^2$. This shows that the level set is a hyperbola.

Chapter 2

2.1 $F_1'(X) = \nabla F_1(X) = (2x_1, -2x_2, 0, 0)$. F_1 has full rank except when $x_1 = x_2 = 0$. $x_1 \neq 0 \Rightarrow \{x_2, x_3, x_4\}$ can be taken as the independent variables, $x_2 \neq 0 \Rightarrow \{x_1, x_3, x_4\}$ can be taken as the independent variables. $F_1(P_1) = -3$. If $x_1 \neq 0$ let $\phi_1(x_2, x_3, x_4) = (x_2^2 - 3)^{1/2}$ near $(x_2, x_3, x_4) = (2, 0, -1)$.
$$F_2'(X) = \begin{pmatrix} 2x_1 & -2x_2 & 0 & 0 \\ 0 & 0 & 2x_3 & -2x_4 \end{pmatrix},$$ full rank $\Leftrightarrow x_1$ or $x_2 \neq 0$ and x_3 or $x_4 \neq 0$. Pairs of independent variables (x_1, x_3), (x_1, x_4), (x_2, x_3), (x_2, x_4). $F_2(P_2) = (1, 3)$, $\phi_2(x_2, x_3) = ((1 + x_2^2)^{1/2}, -(x_3^2 - 3)^{1/2})$ near $(x_2, x_3) = (0, 2)$, $F_3'(X) = \begin{pmatrix} 2x_1 & -2x_2 & 0 & 0 \\ 0 & 0 & 2x_3 & -2x_4 \\ -2x_1 & 0 & 0 & 2x_4 \end{pmatrix}$, full rank $\Leftrightarrow$ any three of the variables x_1, x_2, x_3, x_4 are non-zero. $F_3(P_3) = (-3, -7, 15)$, $\phi_3(x_1) = (\sqrt{x_1^2 + 3}, \sqrt{x_1^2 + 8}, \sqrt{x_1^2 + 15})$ near $X_1 = 1$. Other solutions also exist.

2.2 (i) $x^2 + y^2 = u^2 \cos^2 v + u^2 \sin^2 v = u^2 \Rightarrow u = \pm\sqrt{x^2 + y^2} \Rightarrow \frac{\partial u}{\partial x} = \pm x/\sqrt{x^2 + y^2}$, $y/x = u \sin v / u \cos v = \tan v \Rightarrow v = \tan^{-1}(y/x) \Rightarrow \frac{\partial v}{\partial x} = -y/(x^2 + y^2)$.

    (ii) $\frac{\partial u}{\partial x} \cos v - u \sin v \frac{\partial v}{\partial x} = 1$, $\frac{\partial u}{\partial x} \sin v + u \cos v \frac{\partial v}{\partial x} = 0$. Solving these two linear equations for $\frac{\partial u}{\partial x}$ and $\frac{\partial v}{\partial x}$ gives us $\frac{\partial u}{\partial x} = \cos v$ and $\frac{\partial v}{\partial x} = -\sin v / u$ which agree with (i).

2.3 $F'(X) = \begin{pmatrix} 2x_1 x_2^2 & 2x_1^2 x_2 & 0 & 0 \\ x_2 x_3 & x_1 x_3 & x_1 x_2 & 0 \\ 0 & 0 & 0 & 2x_1 \end{pmatrix}$, $F'(1, 2, 3, 4) = \begin{pmatrix} 8 & 4 & 0 & 0 \\ 6 & 3 & 2 & 0 \\ 0 & 0 & 0 & 8 \end{pmatrix}$, hence $8x_1 + 4x_2 = 0$, $6x_1 + 3x_2 + 2x_3 = 0$, $8x_4 = 0$. Solution set $\{(x_1, -2x_1, 0, 0) : x \in \mathbb{R}\}$, Basis=$\{(1, -2, 0, 0)\}$, Tangent line =$\{(1 + t, 1 - 2t, 3, 4) : t \in \mathbb{R}\}$.

2.4 (a) Let $f(x, y, z) = xe^y - z$. Surface = $f^{-1}(0)$, $\nabla f(x, y, z) = (e^y, xe^y, -1)$, $\nabla f(1, 0, 1) = (1, 0, -1)$, Normal line =$\{(1 + t, t, 1 - t) : t \in \mathbb{R}\}$, Tangent plane =$\{(x, y, z) : (x - 1) \cdot 1 + y \cdot 0 + (z - 1) \cdot (-1) = 0\} = \{(x, y, z) : x - z = 0\}$.

    (b) Let $F(x, y, z) = (x^2 + y^2 - z^2, x + y + z)$. Γ is the set $F^{-1}(1, 5)$. $$F'(x, y, z) = \begin{pmatrix} 2x & 2y & -2z \\ 1 & 1 & 1 \end{pmatrix}, \quad F'(1, 2, 2) = \begin{pmatrix} 2 & 4 & -4 \\ 1 & 1 & 1 \end{pmatrix}$$

    Tangent line = $(1, 2, 2) + \{(x, y, z) : 2x + 4y - 4z = 0, x + y + z = 0\} = (1, 2, 2) + \{(x, y, z) : y = 3z, x = -4z\} = \{(1 - 4t, 2 + 3t, 2 + t) : t \in \mathbb{R}\}$.

2.5 Equation of plane is $ax + by + cz = d$. Using $(1, 2, 3)$ and $(4, 5, 6)$ we obtain $a + 2b + 3c = d$ and $4a + 5b + 6c = d$. Since the plane is perpendicular to the plane $7x + 8y + 9z = 10$ it follows that $(a, b, c) \cdot (7, 8, 9) = 0$, i.e. $7a + 8b + 9c = 0$. Solving for a, b, c and d gives $b = -2a$, $c = a$ and $d = 0$. Solution is $\{(x, y, z) : x - 2y + z = 0\}$.

2.6 Let $f(x, y, z) = \sqrt{x} + \sqrt{y} + \sqrt{z}$, $\nabla f(1, 4, 1) = (1/2, 1/4, 1/2)$. Tangent plane to $f^{-1}(4)$ at $(1, 4, 1)$ is $\{(x, y, z) : (x-1) \cdot (1/2) + (y-4) \cdot (1/4) + (z-1) \cdot (1/2) = 0\} = \{(x, y, z) : x - 2y + z = 8\}$.

2.7 Let $f(x, y, z) = x^2 + 4y^2 + 4z^2$, $\nabla f(x, y, z) = (2x, 8y, 8z)$. Tangent planes to $f^{-1}(1)$ at $(1/\sqrt{2}, 1/4, 1/4)$ and $(\sqrt{3}/2, 0, 1/4)$ are $\sqrt{2}x + 2y + 2z = 2$ and $\sqrt{3}x + 2z = 2$. Line of intersection $= \{(t, (\sqrt{3} - \sqrt{2})t/2, 1 - t\sqrt{3}/2) : t \in \mathbb{R}\}$. $k =$ distance of line to the origin $= (9 - 2\sqrt{6})/(12 - 2\sqrt{6}) = (14 - \sqrt{6})/20$.

2.8 Substitute $x^2 = 1 + y^2$ into $x^2 + 2y^2 = 4$ to get $1 + 3y^2 = 4$. Hence $3y^2 = 3$ and $y = \pm 1$, $x = \pm\sqrt{2}$. Four points are $(\pm\sqrt{2}, \pm 1)$, $(a, b) = (\sqrt{2}, 1)$. Tangent line to hyperbola at $(\sqrt{2}, 1)$, $\sqrt{2}x - y = 1$ has slope $\sqrt{2}$. Normal line to ellipse points in direction $(2\sqrt{2}, 4)$ and hence has slope $4\sqrt{2}/2 = \sqrt{2}$. Both lines pass through $(\sqrt{2}, 1)$ and hence coincide. Area $= 4\sqrt{2}$.

2.9 Direction of normal line to paraboloid at $(1, 1, 4)$ is $(2, 2, -1)$. Tangent plane at $(1, 1, 4)$ is $\{(x, y, z) : 2x + 2y - z = 0\}$. Normal line through $(1, 1, 4)$ is $\{(1 + 2t, 1 + 2t, 4 - t) : t \in \mathbb{R}\}$, $t = -9/8$ gives the point $(-5/4, -5/4, 41/8)$ on paraboloid and on normal. Normal line through $(-5/4, -5/4, 41/8)$ has direction $(-5/2, -5/2, -1)$, $\cos\theta = \frac{(-5/2, -5/2, -1) \cdot (2, 2, -1)}{(25/4 + 25/4 + 1)^{1/2}(4 + 4 + 1)^{1/2}} = \frac{-9}{(27.9/2)^{1/2}} = -\sqrt{\frac{2}{3}}$.

2.10 Let $f(x, y, z) = \log(x^2 + y^2) - 2z$, $S = f^{-1}(0) = \text{Graph}(g)$, where $g(x, y) = \frac{1}{2}\log(x^2 + y^2)$, $\nabla f(1, -1, \frac{1}{2}\log 2) = (1, -1, -2)$. Tangent plane $= \{(x, y, z) : x - y - 2z = 2 - \log 2\}$, Normal line $= \{(1 + t, -1 - t, \frac{1}{2}\log 2 - t) : t \in \mathbb{R}\}$.

2.11 At points of contact normals coincide. Hence $(f'(x), -1, 0) = \lambda(2z, 1, 2z + 2x) \Rightarrow \lambda = -1 \Rightarrow z = -x$ and $f'(x) = (-1)(-2x) \Rightarrow f(x) = x^2$.

Chapter 3

3.1 Let $g(x,y,z) = xy + yz$, $f_1(x,y,z) = x^2 + y^2 - 1$, $f_2(x,y,z) = x - yz$. $\nabla g = \lambda_1 \nabla f_1 + \lambda_2 \nabla f_2 \Rightarrow (y, x+z, y) = \lambda_1(2x, 2y, 0) + \lambda_2(1, -z, -y)$, $y = -\lambda_2 y \Rightarrow y = 0$ or $\lambda_2 = -1$, $y = 0 \Rightarrow x = 0$ which contradicts $x^2 + y^2 = 1$. $\lambda_2 = -1 \Rightarrow x + z = \lambda_1 2y + z \Rightarrow x = \lambda_1 2y = yz \Rightarrow z = 2\lambda_1$, $y = 2\lambda_1 x - 1 = zx - 1 = (x^2/y^2) - 1 \Rightarrow y^2 = x^2 - y = 1 - y^2 - y \Rightarrow 2y^2 + y - 1 = 0$. Hence $y = -1$ or $y = 1/2$, $y = -1 \Rightarrow x = 0$, contradiction. Hence $y = 1/2$, $x = \pm 1/\sqrt{2}$, $z = \pm\sqrt{2}$. Maximum $= 3\sqrt{2}/4$, minimum $= -3\sqrt{2}/4$.

3.2 Maximize z on $\{x^2 + y^2 = 1\} \cap \{x + y + z = 1\}$, $(0,0,1) = \lambda_1(2x, 2y, 0) + \lambda_2(1,1,1) \Rightarrow \lambda_2 = 1$, $2x\lambda_1 = -1 = 2y\lambda_1 \Rightarrow x = y$. $x = \pm 1/\sqrt{2}$, $y = \pm 1/\sqrt{2}$, $z = 1 \pm 1/\sqrt{2} \pm 1/\sqrt{2}$. Maximum $1 + \sqrt{2}$, minimum $1 - \sqrt{2}$.

3.3 Maximize $ab + (1/2)bd$ subject to $b + 2a + 2c = P$, $b^2/4 + d^2 = c^2$. $(b, a + (1/2)d, 0, (1/2)b) = \lambda_1(2,1,2,0) + \lambda_2(0, b/2, -2c, 2d) \Rightarrow c = d$, $b = \sqrt{12}d$, $a = (\sqrt{3} + 1)d \Rightarrow P = (4\sqrt{3} + 6)d$. Maximum $= (2 - \sqrt{3})P^2/4$.

3.4 $(2x, 2y, 2z) = \lambda_1(1,1,-1) + \lambda_2(1,3,1) \Rightarrow 2x = \lambda_1 + \lambda_2$, $2y = \lambda_1 + 3\lambda_2$, $2z = -\lambda_1 + \lambda_2 \Rightarrow 2x - 2y = -2\lambda_2$, $2y + 2z = 4\lambda_2$. If $\lambda_2 = 0$ then $x = y = -z \Rightarrow x = y = 1$, $z = -1$. If $\lambda_2 \neq 0$ then $(x - y)/(y + z) = -1/2 \Rightarrow 2x - y + z = 0 \Rightarrow x = 1$, $y = 1$, $z = -1$. Minimum $= 3$. No maximum distance from the origin.

3.5 $(2x, 2y, 4z) = \lambda_1(1,1,1) + \lambda_2(1,-1,3) \Rightarrow \ldots \Rightarrow x = 17/14$, $y = 16/14$, $z = 9/14$. Minimum $= 707/196$. $z = 3 - x - y = \phi(x,y)$, $f(x,y,\phi(x,y)) = x^2 + y^2 + 2(3 - x - y)^2$ and constraint becomes $2x + 4y = 7$. $\phi_1(x) = (1/4)(7 - 2x)$, $\phi_2(x) = (1/4)(5 - 2x)$, $f(x, \phi_1(x), \phi_2(x)) = x^2 + (7 - 2x)^2/16 + (5 - 2x)^2/8$.

3.6 $(\frac{1}{a}, \frac{1}{b}, \frac{1}{c}) = \lambda(\frac{2x}{a^2}, \frac{2y}{b^2}, \frac{2z}{c^2}) \Rightarrow \lambda \neq 0$, $\frac{1}{a} = \frac{2\lambda x}{a^2} \Rightarrow 1 = \frac{2\lambda x}{a} = \frac{2\lambda y}{b} = \frac{2\lambda z}{c} \Rightarrow \frac{x}{a} = \frac{y}{b} = \frac{z}{c} \Rightarrow x = \pm\frac{a}{\sqrt{3}}$, $y = \pm\frac{b}{\sqrt{3}}$, $z = \pm\frac{c}{\sqrt{3}}$.

3.7 $(yz, xz, xy) = \lambda(-1/x^2, -1/y^2, -1/z^2) \Rightarrow \lambda \neq 0$, $x^2yz = -\lambda = xy^2z = xyz^2 \Rightarrow x = y = z$ and $3/x = 1 \Rightarrow x = y = z = 3$. Maximum 27.

3.8 $V(x,y,z) = xyz$, $(yz, xz, xy) = \lambda(1/a, 1/b, 1/c) \Rightarrow \ldots \Rightarrow x/a = y/b = z/c$. Maximum $= abc/27$.

3.9 Join the vertices to the centre and let x, y, z be the angles at the centre. Products of lengths of sides $= 8R^3 \sin\frac{x}{2} \sin\frac{y}{2} \sin\frac{z}{2} = f(x,y,z)$. Sum of squares of lengths $= 4R^2(\sin^2\frac{x}{2} + \sin^2\frac{y}{2} + \sin^2\frac{z}{2}) = g(x,y,z)$. R is the radius of the circle. $L(x,y,z) = x + y + z - 2\pi$.

(a) $\nabla f = \lambda \nabla h \Rightarrow \cos(x/2) = \cos(y/2) = \cos(z/2) \Rightarrow x = y = z.$

(b) $\nabla g = \lambda \nabla h \Rightarrow \sin(x/2) = \sin(y/2) = \sin(z/2) \Rightarrow x = y = z.$

3.10 Since $x^2 \geq 0$, $3x^2 - y^5 = 0 \Rightarrow y^5 \geq 0 \Rightarrow y \geq 0$. Hence $f(x,y) = 2y \geq 0$ and $f(0,0) = 0$ implies f has minimum value 0 at $(0,0)$. The method of Lagrange multipliers does not work since the surface $3x^2 - y^5 = 0$ does not have full rank at $(0,0)$.

3.11 Minimize $g(\mathbf{v}) = \langle \nabla f(P), \mathbf{v} \rangle$ over $\langle \mathbf{v}, \mathbf{v} \rangle = h(\mathbf{v}) = 1$. $\nabla g(\mathbf{v}) = \lambda \nabla h(\mathbf{v}) \Rightarrow \nabla f(P) = \lambda \mathbf{v}$ and $\|\mathbf{v}\| = 1 \Rightarrow \lambda = \pm \|\nabla f(P)\|/2$. Hence $g(\mathbf{v}) = \ldots = \pm \|\nabla f(P)\|$. Maximum increase in direction $\nabla f(P)/\|\nabla f(P)\|$.

3.12 Let $y_i = x_i/i$. Maximize $n! y_1 \cdots y_n$ on $\sum_{i=1}^{n} y_i^2 = 1$. Use the method in Example 3.2 to get $y_1 = y_2 = \ldots = y_n = 1/\sqrt{n}$ at maximum.

3.13 If x, y and z are perpendicular distances to the sides, of length a, b and c, then it is necessary to maximize d, where $d^2 = x^2 + y^2 + z^2$, subject to constraint Area $= A = (1/2)(ax + by + cz)$. Minimum $= 4A^2/(a^2 + b^2 + c^2)$.

3.14 Nearest point on line, $(11, 2, -4)$. Distance $= \sqrt{6}$.

3.15 Since $-1 \leq \cos\theta \leq +1$, the result in exercise 2.13 is equivalent to the Cauchy–Schwarz inequality. We get the equality case by considering when $\cos\theta = \pm 1$.

Chapter 4

4.1 (a) $\nabla f(x,y) = (2x + y + 2, x + 2) = (0,0) \Rightarrow (-2,2)$ is the only critical point. $H_{f(x,y)} = \begin{pmatrix} 2 & 1 \\ 1 & 0 \end{pmatrix}$, $\det(H_{f(-2,2)}) = -1 < 0 \Rightarrow f$ has a saddle point at $(-2,2)$;

(b) local minimum at $(1,1)$;

(c) saddle point at $(0,1)$;

(d) saddle points at $(2,4)$, $(-1,4)$, $(2,1)$, local maximum at $(1,3)$;

(e) local maxima at $(\pm 1, \pm 1, 1)$, $(\pm 1, 1, \pm 1)$, $(1, \pm 1, \pm 1)$;

(f) critical points at $(\pm 1/\sqrt{2}, \pm 1/\sqrt{2}, \pm 1/\sqrt{2})$, local maxima if even number of negative signs otherwise local minima;

(g) saddle point at $(1, 1, 1/2)$.

4.2 Critical points satisfy $2ax = 2x(ax^2 + by^2 + cz^2)$, $2by = 2y(ax^2 + by^2 + cz^2)$, $2cz = 2z(ax^2 + by^2 + cz^2)$. Critical points are $(0,0,0)$, $(\pm 1, 0, 0)$, $(0, 0, \pm 1)$. If $(x, y, 0)$ is a critical point, $x \neq 0$ and $y \neq 0$ then $2ax/2by = 2x/2y \Rightarrow axy = bxy \Rightarrow a = b$ contradiction. Similarly all critical points can have only one non-zero component. Local minimum at $(0,0,0)$, local maxima at $(\pm 1, 0, 0)$, saddle points at $(0, \pm 1, \pm 1)$.

4.3 Local minimum at $(1/4, 1/4, 1/4)$, degenerate critical points at $(x, 0, 0)$, $(0, y, 0)$, $(0, y, 1 - y)$, $(x, 1 - x, 0)$, $(x, 0, 1 - x)$, $x, y \in \mathbb{R}$.

4.4 The function is only defined when $x \neq 0$, $y \neq 0$ and $z \neq 0$. At a critical point $2x^2yz = y^4z + yz^4 \Rightarrow 2x^3 = y^3 + z^3$. Analogously $2y^3 = x^3 + z^3$ and $2z^3 = x^3 + y^3$. Clearly $(*)\{(x, x, x) : x \neq 0\}$ is a set of critical points. $\frac{2x^3}{2y^3} = \frac{y^3 + z^3}{x^3 + z^3} \Rightarrow x^6 + x^3z^3 = y^6 + y^3z^3 \Rightarrow x^6 - y^6 = y^3z^3 - x^3z^3 \Rightarrow (x^3 - y^3)(x^3 + y^3) = z^3(y^3 - x^3)$. If $x \neq y$ then $x^3 + y^3 = -z^3$. Since $2z^3 = x^3 + y^3$ this implies $z^3 = 0 \Rightarrow z = 0 \Rightarrow 2x^3 = y^3$ and $x^3 = y^3$. This is impossible so $(*)$ gives all critical points.

4.5 Minimize $d^2 = (x + 1)^2 + (y - 1)^2 + (xy - 1)^2$. Critical points satisfy $x + 1 + (xy - 1)y = 0$ and $y - 1 + (xy - 1)x = 0$. Hence $\frac{x+1}{y-1} = \frac{y}{x} \Rightarrow x^2 + x = y^2 - y \rightarrow (x + y)(x - y + 1) = 0 \Rightarrow y = x + 1$ or $y = -x$. If $y = x + 1 \Rightarrow x(x + 1)^2 = 0 \Rightarrow$ solutions $(0, 1)$ or $(-1, 0) \Rightarrow d = \sqrt{2}$. If $y = -x \Rightarrow x^3 + 2x + 1 = 0 \Rightarrow x < 0$ and $d^2 = 2(x + 1)^2 + (x^2 + 1)^2 > 2$. Minimum $= \sqrt{2}$. Lagrange multipliers can also be used for this problem.

4.7 $\nabla\left(\sum_{i=1}^{m} \|X - X_i\|\right) = 2\left(mX - \sum_{i=1}^{m} X_i\right)$.

4.8 $2x + 6\phi\phi_x - 2y - 2y\phi_x = 0$, $4y + 6\phi\phi_y - 2x - 2\phi - 2y\phi_y = 0$ and $\phi_x = \phi_y = 0 \Rightarrow x = y \Rightarrow x = \pm 1$. Critical points of ϕ at $\pm(1,1)$, $\det(H_{\phi(1,1)}) = \det(H_{\phi(-1,-1)}) = 1/4$. Two local minima.

4.9 $AX = \lambda X$, $AY = \mu Y$, $\lambda \neq \mu \Rightarrow \lambda \langle X, Y \rangle = \langle \lambda X, Y \rangle = \langle AX, Y \rangle = \langle X, AY \rangle = \langle X, \mu Y \rangle = \mu \langle X, Y \rangle \Rightarrow (\lambda - \mu)\langle X, Y \rangle = 0 \Rightarrow X \perp Y$ since $\lambda \neq \mu$.

Chapter 5

5.1 $P'(t) = (6\cosh 3t, -6\sinh 3t, 6)$, $\|P'(t)\| = 6\sqrt{2}$, length $= 30\sqrt{2}$.

5.2 (c) On $[0,1]$, $\big(\cos^{-1}(s)\big)' = -(1-s^2)^{1/2}$.

5.3 $s(t) = t\sqrt{r^2 + h^2}$, $P(s^{-1}(t)) = \big(r\cos(\frac{t}{\sqrt{r^2+h^2}}), r\sin(\frac{t}{\sqrt{r^2+h^2}}), \frac{ht}{\sqrt{r^2+h^2}}\big)$.

5.4 **(a)** $s(t) = \sqrt{3}(e^t - 1)$, $s^{-1}(t) = \log(\frac{t}{\sqrt{3}} + 1)$;

    (b) $s^{-1}(t) = \cosh^{-1}(\sqrt{t/2})$.

        Unit speed parametrization $t \to \big(\sqrt{t/2}, \sqrt{1 - (t/2)}, \cosh^{-1}(\sqrt{t/2})\big)$.

5.5 $\big(\sqrt{t^2 - \frac{t^4}{16}}, 4 - \frac{t^2}{4}, t\big)$, $0 \le t \le 4$.

5.6 We need $\phi\colon [0,1] \to [0,1]$, $\phi'(t) > 0$ for $0 < t < 1$, $\phi(0) = 0$, $\phi(1) = 1$, $\phi'(0) = 0$, $\phi'(1) = 0$. Take $\phi(t) = 3t^2 - 2t^3$.

$$P(t) = \begin{cases} (1,2) + (3t^2 - 2t^3)(-2,-4), & 0 \le t \le 1 \\ (-1,-2) + \big(3(t-1)^2 - 2(t-1)^3\big)(5,2), & 1 \le t \le 2 \\ (4,0) + \big(3(t-2)^2 - 2(t-2)^3\big)(-3,2), & 2 \le t \le 3. \end{cases}$$

5.8 This exercise shows that the rate of change of f at X_0 along two curves, which pass through X_0, depends only on the tangents to the curves at X_0.

Chapter 6

6.1 **(a)** $F(P(t)) \cdot P'(t) = t$, $2\pi^2$;

    (b) $F(P(t)) \cdot P'(t) = t^2 + 2t^7 + 3t^6$, $85/84$;

    (c) $F(P(t)) \cdot P'(t) = (\cos e^t, e, e^t) \cdot (0, 1, e^t) = e + e^{2t}$, $(e^8 + 8e - 1)/2$.

6.2 **(a)** and **(b)** do not have a potential;

    (c) $xy + z + \sin(xyz)$;

    (d) $\sin(x^2 + yz)$.

6.4 **(a)** Use Exercise 1.10, $(-2y + z, 2x - 3z, -x + 3y)/\|X\|^3$;

    (b) $(xz^2 - xy^2, yx^2 - yz^2, zy^2 - zx^2)/\|X\|^3$;

    (c) $(0, 0, 0)$.

6.6 Exercises (a) and (c) show that *div* and *curl* behave like derivatives. In (b) the minus sign is, perhaps, unexpected. A very careful application of the definitions is needed to verify these formulae.

6.9 $\nabla(f(\|X\|)) = \nabla(f(\sqrt{x^2 + y^2 + z^2})) = f'(\|X\|)X/\|X\|$, $\nabla^2(f(\|X\|)) = \operatorname{div}(f'(\|X\|)X/\|X\|) = f'(\|X\|)\operatorname{div}(X/\|X\|) + \nabla(f'(\|X\|))X/\|X\| = \ldots = 2f'(\|X\|)/\|X\| + f''(\|X\|)$. g harmonic $\Leftrightarrow \nabla^2(f(\|X\|)) = 0 \Leftrightarrow 2f'(r)/r + f''(r) = 0 \Leftrightarrow r^2 f''(r) + 2r f'(r) = (r^2 f'(r))' = 0 \Leftrightarrow f'(r) = C/r^2 \Leftrightarrow f(r) = B + (A/r)$.

Chapter 7

7.1 $P(t) = (\cos t, 3 \sin t)$, ellipse, maximum curvature at $\pm(3, 0)$.

7.2 $|\kappa| = \left| \dfrac{\frac{2}{a^2}(\frac{2y}{b^2})^2 + \frac{2}{b^2}(\frac{2x}{a^2})^2}{\left((\frac{2x}{a^2})^2 + (\frac{2y}{b^2})^2\right)^{3/2}} \right| = \dfrac{ab}{\left(\frac{x^2 b^2}{a^2} + \frac{a^2 y^2}{b^2}\right)^{3/2}}$. If the ellipse has an anti-
clockwise orientation then $|\kappa| = \kappa$, otherwise $\kappa = -|\kappa|$.

7.4 Let $P(t) = (x(t), y(t), z(t))$ be unit speed. $P'(t) = (x'(t), y'(t), z'(t))$ is
independent of $t \Leftrightarrow$ all tangents are parallel $\Leftrightarrow P(t) = (a + bt, c + dt, e + ft)$.

7.5 (a) $\kappa(t) = (1/4)(2/(1 - t^2))^{1/2}$, $\tau(t) = 1/2\sqrt{1 - t^2}$,

    (b) $\kappa(t) = 1/\sqrt{1 - t^2}$, $\tau(t) = 0$,

    (c) $\kappa(t) = 1/(1 + t^2)$, $\tau(t) = -2/\sqrt{5}(1 + t^2)$.

7.6 $\|P'(t)\| = a \sec \alpha$. Let $ab = \cos \alpha$. Then $Q(t) := (a \cos bt, a \sin bt, t \sin \alpha)$ is
unit speed. $T'(t) = (-ab^2 \cos bt, -ab^2 \sin t, 0)$, $\kappa(t) = ab^2$, $N(t) = (-\cos bt,$
$-\sin bt, 0)$. Centre of curvature $= Q(t) - \kappa(t) N(t) = ((a - ab^2) \cos bt, (a -$
$ab^2) \sin bt, abt \tan \alpha)$, a helix.

7.7 Unit speed parametrization $Q(t) = \left(a \cos(\frac{b}{a} \sinh^{-1}(\frac{t}{b})), a \sin(\frac{b}{a} \sinh^{-1}(\frac{t}{b})), \sqrt{b^2 + t^2} \right)$. Osculating plane at $Q(t)$ is perpendicular to $Q'(t) \times Q''(t)$ and
tangent plane at $Q(t)$ is perpendicular to $\widetilde{Q}(t) = (q_1(t), q_2(t), 0)$. It suffices
to show $\left[Q'(t) \times Q''(t) \cdot \widetilde{Q}(t) / \|Q''(t)\| \cdot \|\widetilde{Q}(t)\| \right] = \alpha(t)$ is constant. $\|\widetilde{Q}(t)\| = a$, $Q'(t) \times Q''(t) \cdot \widetilde{Q}(t) = ab/(b^2 + t^2)$, $\|Q''(t)\| = b\sqrt{a^2 + b^2}/a(t^2 + b^2)$ and
$\alpha(t) = a^3/\sqrt{a^2 + b^2}$.

Chapter 8

8.2 $Q'(t) = B(t)$ implies Q is unit speed and $Q' = T_{\widetilde{\Gamma}} = B$, $Q'' = T'_{\widetilde{\Gamma}} = B' = -\tau N = \kappa_{\widetilde{\Gamma}} N_{\widetilde{\Gamma}}$. Hence $\kappa_{\widetilde{\Gamma}} = \tau$ (since $\tau \geq 0$ and $\kappa_{\widetilde{\Gamma}}$ is positive) and $N_{\widetilde{\Gamma}} = -N$.
$B_{\widetilde{\Gamma}} = T_{\widetilde{\Gamma}} \times N_{\widetilde{\Gamma}} = B \times (-N) = T$, $B'_{\widetilde{\Gamma}} = -\tau_{\widetilde{\Gamma}} N_{\widetilde{\Gamma}} = T' = \kappa N$. Hence $\tau_{\widetilde{\Gamma}} = \kappa$.

8.3 Show $B(t) = (1, -1, 1)/\sqrt{3}$ and use Proposition 8.1. The curve lies in the osculating plane.

8.4 $P'(\theta) = (-\tan\theta, \cot\theta, \sqrt{2})$, $\|P'(\theta)\| = 1/\cos\theta\sin\theta$, $P''(\theta) = (-\sec^2, -\csc^2\theta, 0)$, $\|P'(\theta) \times P''(\theta)\| = \sqrt{2}/(\sin^2\theta\cos^2\theta)$, $\kappa(\theta) = \|P' \times P''\|/\|P'\|^3 = \sin 2\theta/\sqrt{2}$.

8.5 $P'(t) \times P''(t) = 18(2t, 1 - t^2, -1 - t^2)$, $P'''(t) = 6(0, -1, 1)$.

8.6 $\kappa(t) = (\sqrt{2}/3)e^{-t}$, $\tau(t) = (1/3)e^{-t}$.

8.7 The normal plane at $P(t)$ is $\{X : \langle P(t) - X, P'(t) \rangle = 0\}$. If X_0 lies in every normal plane then $\langle P(t) - X_0, P'(t) \rangle = 0$ for all t. (Use example 8.1.) Centre $(-1, 0, 0)$, radius 2.

8.8 $P'(t) = (a, 2bt, 3t^2)$, $P''(t) = (0, 2b, 6t)$, $P'''(t) = (0, 0, 6)$, $P''' \cdot P' \times P'' = 12ab\tau(t)/\kappa(t) = P''' \cdot P' \times P''/\|P' \times P''\|^3 = 12ab(a^2 + 4b^2t^2 + 9t^4)^{3/2}/(36b^2t^4 + 36a^2t^2 + 4a^2b^2)^{3/2}$. P parametrises a generalised helix $\Leftrightarrow a^4 + 4b^2t^2 + 9t^4 = \alpha(36b^2t^4 + 36a^2t^2 + 4a^2b^2)$ for some $\alpha \in \mathbb{R} \Leftrightarrow \alpha = 1/4b^2$ and $4b^2 = \alpha 36a^2 \Leftrightarrow 4b^4 = 9a^2$.

8.9 To show $\langle TX, TY \rangle = \langle X, Y \rangle$ expand $\|T(X + Y)\|^2$ and $\|T(X - Y)\|^2$ and consider the difference. Use exercise 2.13 to show that angles are preserved. To show that area is preserved it suffices (?) to show that the area of rectangles is preserved. (Hint: think of Riemann sums.)

Chapter 9

9.1 $\int_0^1 x^2 dx . \int_0^{\pi/4} \sin^2 y dy = [x^3/3]_0^1 [x/2 - (\sin 2x)/4]_0^{\pi/4} = (\pi - 2)/24.$

9.2 (a) $\int_0^\pi \{\int_0^x x \cos(x + y) dy\} dx = \frac{\pi}{2},$

   (b) $\int_{-1}^{+1} \{\int_{1-\sqrt{(1-x^2)}}^{1+\sqrt{(1-x^2)}} (x^2 + y^2) dy\} dx = \cdots = 8/3 \int_0^1 (1 - x^2)^{1/2} (2 + x^2) dx = \cdots = 3\pi/2;$

   (c) $\int_{1/2}^2 \{\int_{1/x}^2 \frac{y^2}{x^2} dy\} dx = \cdots = 513/192.$

9.3 $\int_0^2 \{\int_{4y-y^2}^{y^2} dx\} dy = 8/3.$

9.4 Let $\mathbf{F} = (Q, -P)$. If $t \to (x(t), y(t))$ is a unit speed parametrization then $\mathbf{n} = (y', -x').$

9.5 (a) $\int_0^2 \{\int_0^{3y} e^{y^2} dx\} dy = 3(e^4 - 1)/2;$

   (b) $\int_0^2 \{\int_{x^2}^{2x} e^{y/x} dy\} dx = (e^2 - 1).$

9.6 $\iint\limits_{\substack{x^2+y^2 \leq 1 \\ x \geq 0, y \geq 0}} (1 - xy) dx\, dy = \int_0^1 \{\int_0^{\sqrt{1-x^2}} (1 - xy) dy\} dx = (2\pi - 1)/8.$ The calculations are easier using polar coordinates (see page 158).

9.8 (a) Once Green's theorem has been used, symmetry implies that the answer is 0.

   (b) By Green's theorem $\int_\Gamma = \iint\limits_{(x-2)^2+y^2<4} 2x dx\, dy.$ This can be evaluated in the usual fashion but some geometry avoids all the calculations: $\int_\Gamma = \iint\limits_{(x-2)^2+y^2<4} 2(x - 2) dx\, dy + 4 \iint\limits_{(x-2)^2+y^2<4} dx\, dy = 0 + 4.2\pi.4 = 32\pi.$ The first integral is zero since $x - 2$ has average value 0 on the disc and the second integral is $4\times$ (area of disc).

   (c) $\int_\Gamma = \iint\limits_\Omega (-3y - 4x^2y) dx\, dy = \int_3^5 dx . \int_1^4 -3y dy - \int_3^5 4x^2 dx . \int_1^4 y dy = -1025.$

Chapter 10

10.1 **(a)** $(u, v) \to (a \cosh u \cos v, b \cosh u \sin v, c \sinh u), -\infty < u < +\infty, 0 < v < 2\pi.$

(b) $(u, v) \to (a \cosh u, b \sinh u \cos v, c \sinh u \sin v), -\infty < u < +\infty, 0 < v < 2\pi;$

10.2 **(c)** $(P_3)_u \times (P_3)_v = (u + v, v - u, -2) \neq (0, 0, 0)$, $P_3(u_1, v_1) = P_3(u_2, v_2) \Rightarrow u_1 + v_1 = u_2 + v_2$ and $u_1 - v_1 = u_2 - v_2 \Rightarrow u_1 - u_2 = v_2 - v_1 = -(v_2 - v_1) \Rightarrow v_1 = v_2$ and $u_1 = u_2.$

10.3 Parametrise the ellipsoid $((x/\sqrt{2}tr)^2 + (y/\sqrt{2}tr)^2 + (z/\sqrt{2}th)^2 = 1$ using ellipsoidal polar coordinates and take the $\theta = \pi/4$ cross section.

10.4 Parametrization formula unchanged. Range $0 < \theta < \pi/2$, $0 < \psi < \pi/2$.

10.5 $\|(0, 0, 1) + t((u, v, 0) - (0, 0, 1))\|^2 = 1 \Rightarrow t = 0$ or $t = 2/(1 + u^2 + v^2)$, $\phi(u, v) = \left(2u/(1 + u^2 + v^2), 2v/(1 + u^2 + v^2), 1 - 2/(1 + u^2 + v^2)\right).$

10.6 Let $P(u) = (u \cos u, u \sin u, u\sqrt{3})$, $0 \leq u \leq 2$, Length $= \int_1^2 (4 + u^2)^{1/2} du = 2(\sqrt{2} + \sinh^{-1}(1))$. (Use the substitution $u = 2 \sinh \theta$.)

10.7 $\nabla f(x, y, z) = (1 + y, x + z, y) \neq (0, 0, 0)$, $U_0 = \{(x, y, z) : y \neq 0\}$ and $\phi_0(x, y) = (x, y, \frac{1 - x - xy}{y})$, $x, y \in \mathbb{R}^2 \setminus \{(x, y); y \neq 0\}$, $U_1 = \{(x, y, z) \in S, y \neq -1\}$ and $\phi_1(y, z) = (\frac{1 - yz}{1 + y}, y, z)$ for $(y, z) \in \mathbb{R}^2 \setminus \{(y, z); y \neq -1\}.$

10.8 At time t, L is at units above the xy-plane. The (x, y) coordinates of a point u units along L are $(u \sin bt, u \cos bt)$ after time t.

Chapter 11

11.1 $P(r,\theta) = (r\cos\theta, r\sin\theta, r^2)$, $0 < r < 2$, $0 < \theta < 2\pi$, $EG - F^2 = r^2(1+4r^2)$, surface area $= \pi(17^{3/2} - 1)/6$.

11.2 **(a)** $P(x,\theta) = (x, x^3\cos\theta, x^3\sin\theta)$, $0 < x < 1$, $0 < \theta < 2\pi$, $EG - F^2 = x^6(1+9x^4)$. Use substitution $u = 1+9x^4$, surface area $= \pi(10^{3/2}-1)/27$

  (b) $P(x,\theta) = (x, x^2\cos\theta, x^2\sin\theta)$, $0 < x < 1$, $0 < \theta < 2\pi$, $EG - F^2 = x^2(1 + 4x^2)$, surface area $= \pi(5^{3/2} - 1)/6$.

11.3 $E = 1$, $F = 0$, $G = r^2 + 1$, surface area $= \pi(\sqrt{2} + \log(1 + \sqrt{2}))$.

11.5 $P(r,\theta) = (r\cos\theta, r\sin\theta, r^3/3)$, $0 < \theta < 2\pi$, $0 < r < \sqrt{3}$. Surface area $= \int_0^{2\pi}\int_0^{\sqrt{3}} r\sqrt{1 + (4r^2/9)}\, dr d\theta = (7\sqrt{21} - 9)\pi/6$.

11.6 Surface Area $= \iint\limits_{x^2+y^2\le a^2} (1 + x^2 + y^2)^{1/2} dx\, dy = \frac{2\pi}{3}((1 + a^2)^{3/2} - 1)$.

11.7 First quadrant $\Rightarrow 0 < \theta < \pi/2$, between the planes $\Rightarrow 0 < r < 8\sqrt{3}$, inside cylinder $\Rightarrow r^2\cos^2\theta < r^2/4 \Rightarrow \cos\theta < \frac{1}{2} \Rightarrow \theta < \pi/3$. $E = (r^2/64) + 1$, $F = 0$, $G = r^2$. Surface area $= \int_0^{\pi/3}\int_0^{8\sqrt{3}} r\sqrt{1 + (r^2/64)}\, dr d\theta = 448\pi/9$.

Chapter 12

12.1 $S \subset \{(x, y, z) : 6x + 3y + 2z = 6\}$, $\mathbf{n}(p) = (6, 3, 2)/7$. $\iint\limits_{S} \mathbf{F} = \frac{18}{7}$(Area S) = 9. (Area of triangle with vertices $\mathbf{a}$, $\mathbf{b}$ and $\mathbf{c}$ is $\frac{1}{2}\|(\mathbf{a}-\mathbf{b}) \times (\mathbf{a}-\mathbf{c})\|$.).)

12.3 $\int_0^1 \int_0^1 (2 + 6u^2v^2 + 2v^9) du\, dv = 43/15$.

12.4 $P(\theta, \psi) = ((b + a\cos\theta)\cos\psi, (b + a\cos\theta)\sin\psi, a\sin\theta)$, $0 < \theta$, $\psi < 2\pi$. $\mathbf{F}(P(\theta, \psi)) = (a\cos\theta\cos\psi, a\cos\theta\sin\psi, a\sin\theta)$. $\langle \mathbf{F}, P_\theta \times P_\psi \rangle = -a^2(b + a\cos\theta)$, $\iint\limits_{S} \mathbf{F} = -4\pi^2 a^2 b$. $\phi_\theta \times \phi_\psi(0, 0) = -a(b + a)(1, 0, 0)$ points inwards, answer $= -(-4\pi^2 a^2 b)$.

12.5 Use spherical polar coordinates (see table 11.1) with range $0 < \theta < \pi/2$, $0 < \psi < 2\pi$. $EG - F^2 = a^4 \sin^2\theta$.

(a) $y^2 + z^2 = a^2 \sin^2\theta \sin^2\psi + a^2 \cos^2\theta$, $4\pi a^4/3$

(b) $(EG - F^2)^{1/2}(x^2 + y^2 + (z + a)^2)^{-1/2} = \sin\theta/(2 + 2\cos\theta)^{1/2}$, $2\pi a(2 - \sqrt{2})$.

12.6 Truncated cone, $f(r, t) = (r\cos\theta, r\sin\theta, r)$, $0 < \theta < 2\pi$, $1 < r < 3$, $f_r \times f_\theta = (-r\cos\theta, -r\sin\theta, r)$, $\langle \mathbf{F}(f(r, \theta)), f_r \times f_\theta \rangle = 2r$, 16π.

12.7 $F_r \times F_\theta = (\sin\theta, -\cos\theta, r)$. $\langle \mathbf{G}(F(r, \theta)), F_r \times F_\theta \rangle == -r\theta \big/ (r^2 + \theta^2)^{3/2}$, $\iint\limits_{S} \mathbf{G} = \int_0^{2\pi} \{\int_\pi^\theta (-r\theta \big/ (r^2 + \theta^2)^{3/2}) dr\} d\theta = \pi(1 + \sqrt{2} - \sqrt{5})$.

Chapter 13

13.1 **(a)** $\mathrm{curl}(z\mathbf{i} - x\mathbf{k}) = (0, 2, 0)$, $(3\pi - 8)2a^2/3$.

     (b) $\mathrm{curl}(-y\mathbf{i} + x\mathbf{j}) = (0, 0, 2)$, πa^2.

     (c) $\mathrm{curl}(-z\mathbf{j} + y\mathbf{k}) = (2, 0, 0)$, $8a^2/3$.

13.2 $P_r \times P_\theta = \frac{r^2}{2b}(\sin\theta + \cos\theta) - r$, $\mathrm{curl}(y, z, x) = (-1, -1, -1)$, $-\pi a^2$.

13.3 Γ lies in sphere centred at the origin, radius $2b$ and is an ellipse with vertices $(b, b, \pm\sqrt{2}b)$, $(2b, 0, 0)$, $(0, 2b, 0)$ and hence a circle of radius $b\sqrt{2}$. Answer $-\sqrt{2}$ (area of circle) $2\sqrt{2}\pi b^2$.

13.6 $P(r, \theta) = (r\cos\theta, r\sin\theta, b(1 - \frac{r\cos\theta}{a}))$, $0 < r < a$, $0 < \theta < 2\pi$ parametrises the portion S of the plane inside the cylinder. $P_r \times P_\theta = (br/a, 0, r)$, $\mathrm{curl}(y - z, z - x, x - y) = (-2, -2, -2)$, $\iint_S \langle \mathrm{curl}(y - z, z - x, x - y), P_r \times P_\theta \rangle dr\, d\theta = -2\pi a(a + b)$. Orientation inconsistent with positive answer. Choose $\theta \to (a\cos\theta, -a\sin\theta, b(1 - \cos\theta))$ to give direction along Γ. Γ is oriented clockwise when viewed from a point far out on the positive z-axis.

Chapter 14

14.1 $P(r, \theta, z) = (r \cos \theta, a + r \sin \theta, z)$, $0 < r < a$, $0 < \theta < 2\pi$, $0 < z < a^2 + r^2 + 2ar \sin \theta/(4a)$, $|det(P')| = r$.

14.2 $f(r, \theta, z) = (r \cos \theta, r \sin \theta, z)$, $0 < \theta < \pi$, $0 < z < r$, $0 < r < 2a \sin \theta$, volume$= \int_0^\pi \{\int_0^{2a \sin \theta} \{\int_0^r r dz\} dr\} d\theta = 8a^3/3 \int_0^\pi \sin \theta (1 - \cos^2 \theta) d\theta = 32a^3/9$.

14.3 Use $(r, \theta, z) \to (r \cos \theta, \sin \theta, z)$, $0 < r < 1$, $0 < \theta < 2\pi$, $0 < z < 1$.

14.5 (a) 8π;

   (b) 8π;

   (c) $\frac{81\pi}{4} - \frac{266}{5}$.

14.8 $V = \int_0^1 \{\int_0^1 \{\int_0^{(1+x+y)^{1/2}} dx\} dy\} dx = 4(3^{5/2} - 2^{7/2} + 1)/15$, by symmetry the second volume $= 8V$.

14.9 A quarter of an ice-cream cone.

14.11 $\iiint_V x dx\, dy = \int_0^1 u^3 du \int_0^1 v(1 - v) dv \int_0^1 dw = 1/24$,

   $\iiint_V \frac{dx\, dy\, dz}{y+z} = \int_0^1 \int_0^1 \int_0^1 \frac{u^2 v}{uv} du\, dv\, dw = 1/2$.

Chapter 15

15.2 $\operatorname{div}(\mathbf{F}) = y^2 + x^2$. Interior of $S = \{(r\cos\theta, r\sin\theta, z) : 0 < r < \sqrt{3}, \pi/2 < \theta < 3\pi/2, 0 < z < -r\cos\theta\}$, $\operatorname{div}(\mathbf{F}) = r^2$, $18\sqrt{3}/5$;

15.3 $\int_0^a (\int_0^{b(1-\frac{x}{a})} (\int_0^{c(1-\frac{x}{a}-\frac{y}{b})} x\,dz)\,dy)\,dx = a^2cb/24$.

15.4 Order of integration is important, first x and then y to get $(e-2)/2e$.

15.5 $\operatorname{div}(x^2, -2xy, 3xz) = 5x - 2y$.

$$\iiint\limits_{\substack{x,y,x\geq 0 \\ x^2+y^2+x^2\leq 4}} x\,dx\,dy\,dz = \iiint\limits_{\substack{x,y,x\geq 0 \\ x^2+y^2+x^2\leq 4}} y\,dx\,dy\,dz = \pi.\ 3\pi.$$

Chapter 16

16.1 $E = 1$, $F = 0$, $G = t^2 + b^2$, $l = 0$, $m = -b(t^2 + b^2)^{-1/2}$,

$K = -b^2 / (t^2 + b^2)^2$.

16.2 $P(r, \theta) = (r \cos\theta, r \sin\theta, r)$, $\mathbf{n} = (-\cos\theta, -\sin\theta, 1)/\sqrt{2}$, $K = 0$.

16.3 $P = (1, 1, 2)$, $P'(1)/\|P'(1)\| = (2, 1, 6)/\sqrt{41} = \mathbf{v}$, $S = g^{-1}(0)$ where
$g(u, v, w) = u^2 + v^2 - w$, $\nabla g/\|\nabla g\| = \frac{(2u, 2v, -1)}{(4u^2 + 4v^2 + 1)^{1/2}}$, $\frac{\nabla g}{\|\nabla g\|}$ coincides with
$\mathbf{n}$ on S and is defined on an open set containing S. Calculate $D(\nabla g/\|\nabla g\|)$
and then $\kappa_P(\mathbf{v}) = -D_\mathbf{v}(\nabla g/\|\nabla g\|) = -D(\nabla g/\|\nabla g\|) \cdot \mathbf{v} = \ldots = 634/1107$.

16.5 $K = 36uv \cdot (1 + 9u^4 + 9v^4)^{-2}$, elliptic points $u > 0$, $v > 0$, hyperbolic points
when $uv < 0$.

16.6 $l = m \equiv 0 \Rightarrow K \equiv 0$.

16.7 $u \to (av, -bv, 0) + u(a, b, v)$, $v \to (au, bu, 0) + v(a, -b, u)$.

16.8 $K = \frac{1}{a^2 b^2 c^2} \left(\frac{x^2}{a^4} + \frac{y^2}{b^4} + \frac{z^2}{c^4} \right)^{-2}$.

16.9 $P(x, y) = (x, y, x^2 - y^2)$, $\mathbf{n} = (-2x, -2y, 1) / (1 + 4x^2 + 4y^2)^{1/2}$, $\mathbf{n}(p) = (0, 0, 1)$. Let $\mathbf{v} = (v_1, v_2, v_3)$. By Exercise 16.4, $\langle L_P(\mathbf{v}), \mathbf{v} \rangle = v_1^2 - v_2^2 = k_p(\mathbf{v}) \kappa_p(1/\sqrt{2}, 1/\sqrt{2}, 0) = 0$, $(t/\sqrt{2}, t/\sqrt{2}, 0)$ lies in the surface for all t.

16.10 By inspection, $f(x_1, x_2, x_3, x_4) = 6x_1^2 + 3x_2^2 + 6x_3^2 + 11x_4^2 - 4x_1 x_2 - 2x_1 x_3 + 2x_1 x_4 - 4x_2 x_3 - 2x_3 x_4$. Local minimum.

16.11 By using the mixed terms, e.g. $x_i x_j$ with $i \neq j$, $f(x_1, x_2, x_3, x_4) = (2x_1 - x_2)^2 + (x_1 - x_3)^2 + (x_2 + x_4)^2 + (x_2 - 2x_3)^2 + (x_2 + 3x_4)^2 + (x_3 - x_4)^2$.
This also shows that f has a local minimum at the origin. One can always
write a quadratic form in n variables as $\sum_{i=1}^{n} \lambda_i g_i^2$ where $\lambda_1, \ldots, \lambda_n$ are
the eigenvalues of $H_{f(0,0)}$ and g_i are linear.

Chapter 17

17.1 Use toroidal polar coordinates and Proposition 17.1, $L(\alpha P_\theta + \beta P_\psi) = \frac{\alpha}{a} P_\theta + \frac{\beta \cos \theta}{b + a \cos \theta} P_\psi$.

17.2 $\phi(u,v) = (u, v, u^2 - 2v^2)$, $\kappa_P\left(\frac{v_1 \phi_u + v_2 \phi_v}{\|v_1 \phi_u + v_2 \phi_v\|}\right) = \frac{2v_1^2 - 4v_2^2}{(4u^2 + 16v^2 + 1)^{1/2} \|v_1 \phi_u + v_2 \phi_v\|^2}$
when $P = \phi(u,v)$, $t \to \phi(\sqrt{2}t, t) = (\sqrt{2}t, t, 0)$.

17.3 Suppose $\kappa \neq 0$ and $\mathbf{n}_1$ and $\mathbf{n}_2$ (the normals to $\mathbf{S}_1$ and $\mathbf{S}_2$) are linearly independent. $\|\mathbf{n}_2 - \langle \mathbf{n}_1, \mathbf{n}_2 \rangle \mathbf{n}_1\| = \sin^2 \theta$. If T is the unit tangent to Γ and N is the normal in $\mathbb{R}^3$ then $T \perp \mathbf{n}_1, \mathbf{n}_2$ and $N \in \mathrm{span}(\mathbf{n}_1, \mathbf{n}_2)$. Hence $N = \langle N, \mathbf{n}_1 \rangle \mathbf{n}_1 + \frac{\langle N, \mathbf{n}_2 - \cos \theta \mathbf{n}_1 \rangle}{\sin \theta} \frac{(\mathbf{n}_2 - \cos \theta \mathbf{n}_1)}{\sin \theta}$. On S_i, $\lambda_i = k_P(T) = -\langle D_T \mathbf{n}_i, T \rangle = \langle \mathbf{n}_i, D_T T \rangle = \langle \mathbf{n}_i, \kappa N \rangle$, $\kappa N = \lambda_1 \mathbf{n}_1 + \frac{\lambda_2 - \cos \theta \lambda_1}{\sin \theta} \frac{(\mathbf{n}_2 - \cos \theta \mathbf{n}_1)}{\sin \theta}$, $\kappa^2 = \lambda_1^2 + \frac{\lambda_2^2 - 2 \cos \theta \lambda_1 \lambda_2 + \cos^2 \theta \lambda_1^2}{\sin^2 \theta}$, $\kappa^2 \sin^2 \theta = \lambda_1^2 \sin^2 \theta + \lambda_2^2 - 2 \cos \theta \lambda_1 \lambda_2 + \cos^2 \theta \lambda_1^2 = \lambda_1^2 + \lambda_2^2 - 2 \cos \theta \lambda_1 \lambda_2$. $\kappa = 0 \Rightarrow \lambda_1 = \lambda_2 = 0$. $\mathbf{n}_1$ and $\mathbf{n}_2$ are not linearly independent $\Rightarrow \mathbf{n}_1 = \pm \mathbf{n}_2$ and $\lambda_1 = \pm \lambda_2$.

17.4 By Exercise 16.4, $\kappa_P(\mathbf{v}) = \frac{v_1^2 l + 2 v_1 v_2 m + v_2^2 n}{v_1^2 E + 2 v_1 v_2 F + v_2^2 G} = \alpha$ (constant) for all $(v_1, v_2) \Leftrightarrow v_1^2 (l - \alpha E) + 2 v_1 v_2 (m - \alpha F) + v_2^2 (n - \alpha G) = 0$ for all $(v_1, v_2) \Leftrightarrow l = \alpha E$, $m = \alpha F$ and $n = \alpha G$. By Example 16.1 and the above $z = xy$ has no umbilics.

17.5 Use Euler's formula.

17.6 $P(x,y) = (x, y, \log \cos y - \log \cos x)$, $E = \sec^2 x$, $F = -\tan x \tan y$, $G = \sec^2 y$, $l = (1 + \tan^2 x) / (1 + \tan^2 x + \tan^2 y)^{1/2}$, $m = 0$, $n = -(1 + \tan^2 y) / (1 + \tan^2 x + \tan^2 y)^{1/2}$.

Chapter 18

18.1 Cylinder $= f^{-1}(0)$, $f(x, y, z) = x^2 + y^2 - 1$, $\nabla f = (2x, 2y, 0)$. Hence $\mathbf{n}(\phi(t)) \parallel (2\cos(at+b), 2\sin(at+b), 0)$, $\phi''(t) = (a^2\cos(at+b), a^2\sin(at+b), 0)$

18.2 Unit speed parametrization $\phi(t) = (x(t), y(t)) \Rightarrow x''x' + y''y' = 0$.

$\mathbf{n} \parallel (-y', -x'\cos, -x'\sin)$, $\phi'' = (x'', y''\cos, y''\sin) \Rightarrow \phi'' \parallel \mathbf{n}$.

18.3 $P = (x_0, y_0, z_0)$, $P + t(a, b, c)$ lies in the surface $\Leftrightarrow (x_0 + ta)^2 - (y_0 + tb)^2 = z_0 + tc$ for all $t \Leftrightarrow (2x_0a - 2y_0b - c)t + (a^2 - b^2)t^2 = 0$ for all t. Letting $a = 1$, $b = 1$, $2x_0 - 2y_0 = c$ and $a = 1$, $b = -1$, $2x_0 + 2y_0 = c$ gives two lines on the surface.

18.4 Let P be a unit speed parametrization of Γ and $T(t) = P'(t)$, $\kappa_n(t) = \kappa_{P(t)}(T(t))$, $\tau_g = 0 \Leftrightarrow (\mathbf{n})' = -\kappa_n T \Leftrightarrow L_{P(t)}(T(t)) = \kappa_n(t)T(t) \Leftrightarrow T(t)$ is a principal curvature for all $t \Leftrightarrow \Gamma$ is a line of curvature, $\kappa_g = 0 \Leftrightarrow N = \pm \mathbf{n}$. If $N = \mathbf{n}$ then $\kappa_n = \kappa$ and $B = T \times N = T \times \mathbf{n} = -\mathbf{n}_s$. If $N = -\mathbf{n}$ then $\kappa_n = -\kappa$ and $B = T \times N = T \times (-\mathbf{n}) = \mathbf{n}_s$. $N' = \kappa T + \tau B$. If $N = \mathbf{n}$ then $-\kappa T + \tau B = -k_n T - \tau_g(-B) \Rightarrow \tau = \tau_g$. If $N = -\mathbf{n}$ then $-\kappa T + \tau B = k_n T + \tau_g B \Rightarrow \tau = \tau_g$.

18.5 The normal curvature of a sphere curve $= \pm 1/r$ ($r = $ radius of the sphere). This is geometrically obvious and can be proved mathematically using spherical polar coordinates (Table 11.1) and Exercise 16.4. Hence κ is constant $\Leftrightarrow \kappa_g$ is constant. If $\tau \neq 0$ then, by Example 8.1, $\langle P(t) - c, B(t) \rangle = -\frac{1}{\tau}(\frac{1}{\kappa})' = 0$ and $P(t) - c = \frac{1}{\kappa}N(t)$. Hence $\kappa = 1/r$ and $\kappa_g = 0$. By Example 18.1 the curve lies in a circle. If $\tau \equiv 0$ then κ is constant, by Example 18.1, and by Proposition 8.1 the curve is a plane curve of constant curvature and hence a circle. Of course we may have τ zero at an isolated point (or even on a convergent sequence of points). More delicate analysis is required for this case. In such problems the case where $\tau \neq 0$ should be considered initially. This usually exposes the geometry of the situation. The $\tau \equiv 0$ case reduces to a problem of plane curves (Proposition 8.1) and is usually easy. These two cases can often be combined to give the general solution. Similar remarks apply to curvature in $\mathbb{R}^n$ (i.e. to the cases $\kappa > 0$ everywhere, and $\kappa \equiv 0$).

18.6 $\kappa_{P(t)}(T(t)) = 0 \Leftrightarrow N \perp \mathbf{n} \Leftrightarrow N$ is a tangent vector to surface. Since $T(t)$ is also a tangent vector to the surface we have osculating plane $=$ span $(T(t), N(t)) = $ tangent plane to surface $\Leftrightarrow N(t)$ is tangent to the surface.